Flashback Mechanisms in Lean Premixed Gas Turbine Combustion

Flashback Mechanisms in Lean Premixed Gas Turbine Combustion

Ali Cemal Benim
Khawar J. Syed

AMSTERDAM • BOSTON • HEIDELBERG • LONDON
NEW YORK • OXFORD • PARIS • SAN DIEGO
SAN FRANCISCO • SINGAPORE • SYDNEY • TOKYO
Academic Press is an imprint of Elsevier

Academic Press is an imprint of Elsevier
225 Wyman Street, Waltham, MA 02451, USA
525 B Street, Suite 1800, San Diego, CA 92101-4495, USA
32 Jamestown Road, London NW1 7BY, UK
The Boulevard, Langford Lane, Kidlington, Oxford OX5 1GB, UK

Notices
Knowledge and best practice in this field are constantly changing. As new research and experience broaden our understanding, changes in research methods, professional practices, or medical treatment may become necessary.

Practitioners and researchers must always rely on their own experience and knowledge in evaluating and using any information, methods, compounds, or experiments described herein. In using such information or methods they should be mindful of their own safety and the safety of others, including parties for whom they have a professional responsibility.

To the fullest extent of the law, neither the Publisher nor the authors, contributors, or editors, assume any liability for any injury and/or damage to persons or property as a matter of products liability, negligence or otherwise, or from any use or operation of any methods, products, instructions, or ideas contained in the material herein.

Library of Congress Cataloging-in-Publication Data
A catalog record for this book is available from the Library of Congress

British Library Cataloguing-in-Publication Data
A catalogue record for this book is available from the British Library

ISBN: 978-0-12-800755-6

For information on all Academic Press publications
visit our website at http://store.elsevier.com/

Typeset by Thomson Digital

CONTENTS

AUTHOR BIOS

Professor Dr.-Ing. Ali Cemal Benim received his B.Sc. and M.Sc. degrees in Mechanical Engineering at the Bosphorus University, Istanbul, Turkey. The topic of his Master's Thesis was "Finite Element Solution of Navier–Stokes Equations." He received his Ph.D. degree at the University of Stuttgart, Germany, in 1988, on the topic "Finite Element Modeling of Turbulent Diffusion Flames" with "Degree of Distinction." During a postdoctoral period at the University of Stuttgart, he performed research on modeling pulverized coal combustion. In 1990 he joined ABB Turbo Systems Ltd. in Baden, Switzerland, where his activities were focused on gas turbine combustion. He was the manager of the "Computational Flow and Combustion Modeling" group. Since January 1996, he has been Professor for Energy Technology at the Düsseldorf University of Applied Sciences, Germany. He authored over 150 publications in a number of fields including: finite element methods in flow problems, the finite analytic method, Navier–Stokes solution techniques, the lattice Boltzmann method, turbulence modeling, combustion modeling (gaseous, liquid, and pulverized fuels), gasification, convective and radiative heat transfer, metal flow with phase change, two-phase flows, erosion, turbomachinery, internal and external aerodynamics, thermoelectricity, and biofluid dynamics. He holds several patents on gas turbine cooling. Prof. Benim is the Executive Editor of *Progress in Computational Fluid Dynamics – An International Journal* and has editorial positions in a number of further international journals.

Dr. Khawar Jamil Syed received his M.Sc. degree in 1984 in Mechanical Engineering, for which his research focused on the liftoff stability of turbulent nonpremixed flames. He gained his Ph.D. degree on the topic "Soot and Radiation Modelling in Buoyant Fires" in 1990. Both degrees were received at Cranfield University, UK, where he also continued as a Post-Doctoral Research Fellow, researching and lecturing in combustion modeling. He continues to visit Cranfield University on a regular basis where he contributes to the annual Gas Turbine Combustion Short Course. In 1991 he joined ABB Power Generation Ltd. in Baden, Switzerland, as a Senior Development Engineer, where he was in charge of the combustion modeling activity. Between 1998 and 2007 he worked at Siemens Industrial Turbomachinery Ltd. in Lincoln, UK, where he held several positions: Combustion Technology Group Leader, Combustion Technology Manager, Engineering Methods Manager, and Combustor Component Owner. Since 2007 he has been the Group Manager for Combustor Technology at Alstom (Switzerland) Ltd., being responsible for the development of combustor technologies to be implemented into future gas turbine products. Since 2012 he holds an Honorary Visiting Professorship at Cardiff University, UK. He has authored about 30 papers on combustion modeling and gas turbine combustor technology and holds several patents in the area.

PREFACE

The purpose of the present book is to provide an up-to-date review of flame flashback phenomena for lean premixed burners. The importance of this topic has grown in recent times, due to both increasing firing temperatures of land-based gas turbines and growing interest in alternative fuels that contain hydrogen, which are more reactive than natural gas.

Many publications exist on the study of flashback, which focus on different flashback mechanisms. This book collates these works and structures their key findings into the different modes of flashback that can result. The key physical phenomena are described as are the parameters that influence them.

Ali Cemal Benim and Khawar Jamil Syed

CHAPTER 1

Introduction

In industrial gas turbines, to achieve low pollutant emissions, a variety of preformation and postformation control technologies are employed. Among the former, "lean premixed combustion" is now the standard for achieving the low NO_X emission targets that are in place for natural gas applications (Eroglu et al., 2001; Krebs et al., 2010; Sattelmayer, 2010).

The control principle utilized by lean premixed technology is the strong temperature dependence of the NO formation rate. The low flame temperatures attained under lean conditions effectively limit NO formation. For nonpremixed combustion, however, even if the fuel and air are fed "globally" in a lean ratio into the combustion chamber, the whole range of equivalence ratios occur "locally" within the combustion zone, leading to reactions occurring around stoichiometric conditions, and, thus, to locally high temperatures and high NO formation rates. Only after complete and uniform mixing of fuel with air, locally lean conditions can be achieved throughout the combustion zone, and the potential of lean combustion can be fully exploited. On the other hand, to accomplish a satisfactory degree of mixing within the limited space and residence time available in the premixing section of a gas turbine burner is not straightforward. Difficulties arise because of conflicting requirements from different design criteria such as mixing, pressure drop, robustness, reliability, and flashback safety, the latter being the focus of the present treatise.

Flashback is the undesirable penetration of flame into the burner, resulting in combustion within the premixing section. Indeed, a tendency toward flashback is an intrinsic feature of all premixed combustion systems, including gas turbine combustors (Lefebvre, 1983), as, in such systems, a burnable mixture is always available upstream the combustor. Flashback can take place, for example, when the local flame speed exceeds the approach flow velocity, in all regions, where fuel/air mixture exists in flammable proportions.

Flashback Mechanisms in Lean Premixed Gas Turbine Combustion

Burners for land-based gas turbines are designed to predominantly accept natural gas as fuel. There is also a requirement for the same hardware to accept liquid fuel, typically diesel oil. Depending on the burner technology, low NO_X emissions may be achieved without water injection, where the liquid burns in a premix mode or NO_X may be controlled by injecting water to reduce flame zone temperatures.

However, interest in utilizing other energy resources, due to concerns about the environment and energy security, has stimulated the investigation of alternative fuels, such as coal-derived syngas, biomass, landfill gas, or process gas (Lieuwen et al., 2008a) possibly in combination with technologies such as integrated gasification combined cycle (IGCC). Such fuels can contain large amounts of hydrogen, which is significantly more reactive than natural gas, and therefore have greater flashback propensity. Hydrogen has a laminar flame speed, which is about four times that of typical natural gases. Its turbulent flame speed and its resistance to hydrodynamic strain are also greater. Consequently, the risk of flashback grows and its abatement becomes a significant challenge.

Even if a burner is designed to operate safely for a design point, flashback can still occur if the flow field of the burner temporarily experiences deviations from the design, which can be caused by some kind of disturbance, for example, rapid deloading of the gas turbine. This transience may lead to momentarily very low flow velocities and therefore gives the opportunity for the flame to enter deep into the burner and ignite fuel within the premixing zone, for example, in the wakes of the fuel injection jets. In these wakes, the fuel concentration exhibits values around stoichiometric, implying much higher laminar flame speeds compared with the well-mixed downstream region. The comparably high local turbulence intensities may additionally increase the local turbulent flame speed. These features favor an anchoring of the flame after such a transience has occurred. If the burner is robust, when this transience is over, the flame should be extinguished in these zones and move into the combustor. If the burner is not robust, when the transience is over, the flame does not extinguish and remains in the burner. In this case NO_X emissions will be high and burner damage may occur. This propensity for flame anchoring in low-velocity regions in the burner, after a temporary flashback, is usually referred to as the "flameholding" behavior of the burner.

Thus, flashback can be viewed from two perspectives, both of which must be addressed to ensure a flashback-safe design. One is to ensure that, under steady conditions, the flame speed nowhere exceeds flow velocities. The critical phenomenon in this case is flame propagation. The other assumes that the flame can transiently move into the burner, and it must be ensured that the flame must extinguish in the burner, once the transience is over. In this case the critical phenomenon is extinction. In the present essay, both perspectives are addressed.

CHAPTER 2

Concepts Related to Combustion and Flow in Premix Burners

2.1 LAMINAR PREMIXED FLAMES

2.1.1 The Laminar Flame Speed

The laminar flame speed is the speed at which a flame will propagate through a quiescent, homogeneous mixture of unburned reactants, under adiabatic conditions (Turns, 2012). The laminar flame speed for a planar, unstretched flame (S_L^0) shall be dealt with first. Consider a one-dimensional, planar flame front within the laminar flow of a homogenous mixture, where the unburned reactants approach the flame front with the constant velocity u. Steady-state conditions, i.e., a spatially stationary flame front is obtained, when $u = S_L^0$. In the following, as is common practice, a number of assumptions are made. It is assumed that the Mach number is low, and mechanical energies, the viscous dissipation, as well as the pressure difference across the flame front are negligible. Furthermore, it is assumed that the specific heat capacity, the thermal conductivity, and the diffusivity take constant values across the flame front (values corresponding to the unburned mixture are taken), the flame is thin (high activation energy), the Lewis number is unity (Le = 1), and the chemical kinetics is governed by a single-step irreversible reaction.

The initial theoretical analyses for the determination of the laminar flame speed date back to Mallant and Le Chatelier, who postulated that the combustion is sustained, i.e., the unburned mixture is continuously heated up to the "ignition temperature," by the upstream propagation of heat through the layers of unburned gas (Glassmann and Yetter, 2008). Here, the flame is assumed to consist of two zones, i.e., a "preheat zone" and a "reaction zone," as illustrated in Figure 2.1 for a one-dimensional flame.

Expressions derived based on simplifying assumptions indicate that S_L^0 strongly depends on T_U and T_B. Among the improvements of this

Flashback Mechanisms in Lean Premixed Gas Turbine Combustion

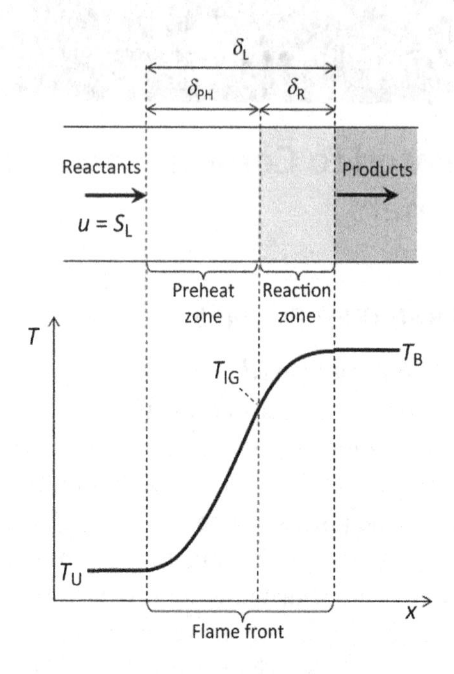

Fig. 2.1. Schematic of a one-dimensional planar and unstretched flame front.

theory, that postulated by Zeldovich and Frank-Kamanetskii has been most significant (Glassmann and Yetter, 2008), who considered the diffusion of molecules. In addition to heat diffusing from the reaction zone into the preheat zone, the reactants diffuse, in the opposite direction, from the preheat zone into the reaction zone, which is equally important for sustaining the combustion. The relative importance of both mechanisms is indicated by the Lewis number:

$$\mathrm{Le} = \frac{\alpha}{D} \tag{2.1}$$

In the theories outlined earlier, a problematical quantity is the ignition temperature, which cannot easily be determined. There are less sophisticated approaches that still lead to some useful estimations. The assumption of a linear temperature profile across the whole flame front, along with the assumption of equal sizes for the preheat and reaction zones, implies an ignition temperature that is equal to the arithmetic average of the unburned and burned mixture temperatures. Equating the heat conducted from the reaction zone (which is now straightforward based on the assumed linear temperature profile and Fourier's law) to the energy required to raise the temperature of the unburned gases (that flow toward the flame front) to the ignition temperature, and considering the equality of flow and flame speeds, for a stationary flame, the following relationship between the laminar flame speed and flame thickness is obtained (Turns, 2012):

$$S_L = \frac{\alpha}{\delta_{PH}} = \frac{\alpha}{\delta_R} = \frac{2\alpha}{\delta_L} \tag{2.2}$$

The equality $\delta_{PL} = \delta_R$ is, of course, a strong simplification. Nevertheless, the proportionality $\delta \sim \alpha / S_L$ is an important correlation, which is quite often used as equality $\delta = \alpha / S_L$ for estimating orders of magnitudes.

An analysis of the energy balance assuming single-step kinetics enables the establishment of a proportionality between the unburned reactant temperature T_U, burned product (adiabatic flame) temperature T_B (which, in turns, depends on the fuel composition, the unburned mixture temperature, and the equivalence ratio), the pressure p, and the laminar flame speed S_L (Turns, 2012), given as follows:

$$S_L \sim \left(\frac{T_U + T_B}{2} \right)^{0.375} \cdot T_U \cdot T_B^{-(n/2)} \cdot \exp\left(-\frac{E_A}{2RT_B} \right) \cdot p^{(n-2)/2} \tag{2.3}$$

As can be seen from the above equation, the temperature influence is governed by the exponential term. The pressure influence depends on the overall reaction order n. The overall reaction order can be determined experimentally or by numerical analysis based on detailed reaction mechanisms. Strictly speaking, the overall reaction order is not constant and depends on further variables such as equivalence ratio and pressure (Law, 2006). For single-step kinetics of methane, Westbrook and Dryer (1981) give $n = 1$, which results in $S_L \sim p^{-0.5}$. This value, which may be considered to loosely correspond to the ranges provided in Law (2006), is widely assumed and generally used for methane (Turns, 2012). For higher hydrocarbons, n is generally assumed to take values smaller than 2 (Peters and Rogg, 1993), implying a general decrease of the laminar flame speed with pressure. For hydrogen–air flames, according to the analysis of Law (2006), n shows even stronger variations with pressure, where under lean conditions, for pressures lower than 50 atm, values lower than 1 are indicated.

Based on the flame thickness and speed, a chemical timescale τ_C can be defined shown in the following. This corresponds to a residence time in the flame zone, and can also be written in terms of ν, assuming Pr = 1, as follows:

$$\tau_C = \frac{\delta_L}{S_L} = \frac{\alpha}{S_L^2} \approx \frac{\nu}{S_L^2} \approx \frac{\delta_L^2}{\nu} \tag{2.4}$$

The previous expressions assume, among other things, a Lewis number of unity. Beyond the discussion of the mixture Lewis number being unity or not, the calculation of a meaningful effective Lewis number can further be complicated, if the individual unburned species exhibit considerably different mass diffusivities. This phenomenon is generally referred to as "preferential diffusion." Hydrogen fuel blends typically exhibit this behavior. Since hydrogen possesses a comparably much larger mass diffusivity, it can diffuse much more rapidly to the flame zone, leading to a local hydrogen enrichment and shifting of the local mixture composition.

Beyond the analytical theories, a one-dimensional laminar premixed flame can be analyzed in detail by solving the complete set of transport equations numerically, using a chemical simulation tool such as CHEMKIN (www.reactiondesign.com) or Cantera (www.cantera.org), utilizing a detailed reaction mechanism such as GRI-Mech (www.me.berkeley.edu/gri-mech). This leads to an accurate prediction of the laminar flame speed, considering the species diffusion accurately, including Lewis number effects and the preferential diffusion. However, if (Eq. 2.4) is then used to estimate the chemical timescale, it should be recalled that the expression still assumes a Lewis number of unity.

2.1.2 Effect of Flame Curvature and Stretch on the Laminar Flame Speed

A planar and unstretched flame front is an idealization. In most real cases, the flame front is curved/wrinkled and stretched. This is especially true for turbulent flames, due to the action of turbulent eddies. Curvature and stretch can affect the flame speed. A stretching of the flame occurs if the points on the flame surface "glide" along it due to a tangential velocity component that exhibits a spatial variation. Thus, a Lagrangian flame surface area A spanned by a number of points is continuously deformed. The flame stretch rate is defined as follows:

$$K = \frac{1}{A}\frac{dA}{dt} \tag{2.5}$$

Consider a curved flame surface that is convex with respect to the unburned mixture (Figure 2.2).

As the unburned gas flows against this surface, the streamlines diverge and the resulting tangential velocity positively stretches the surface. For a Eulerian control volume spanned along a portion of the flame front (as indicated in Figure 2.2), positive stretch means that mass flow parallel to the flame surface, created by the tangential flame velocity, is larger for the outflow compared with that for the inflow, so that there is a net outflow. This causes the heat transferred from the reaction zone into the preheat zone to be transported out of the control volume and then it cannot contribute to preheating of the unburned mixture. This heat loss tends to reduce the local temperatures and, thus, to decrease the laminar flame speed. In addition to flame stretch, curvature also plays a role. One can see that the heat flux into the preheat zone is divergent due to the curvature, which also tends to reduce the local temperatures and decrease the flame speed around the flame tip. On the contrary, the diffusion of reactants to the flame surface is convergent, which, in turn, tends to increase the propagation velocity. The net response of these two diffusive fluxes to flame curvature depends on the Lewis number. For Le > 1,

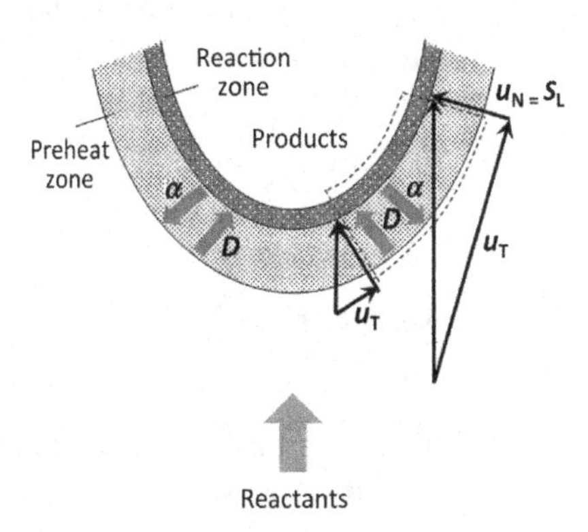

Fig. 2.2. Schematic of a curved and stretched flame front; u_N, velocity component normal to flame surface; u_T, velocity component tangential to flame surface.

for a flame surface curved convex to reactants (Figure 2.2) the net effect of heat and mass diffusion will be a reduction in the flame speed. In parts concave to reactants (the reverse case compared with Figure 2.2), the flame speed will increase. This situation (Le > 1) is regarded as stable, since by the reduction/increase of the flame speed in parts convex/concave toward the reactants, the flame front will straighten out. On the contrary, for Le < 1, the flame speed on surfaces convex/concave toward the reactants will increase/decrease, which will lead to a growth of the curvature. Thus, this situation is regarded as unstable, leading to a self-induced wrinkling of an otherwise planar laminar flame. For most hydrocarbon–air mixtures, Lewis numbers are quite close to unity. On the contrary, lean hydrogen–air mixtures exhibit Lewis numbers that are significantly less than unity leading to unstable flames in the sense described earlier.

The response of the flame to stretch is characterized quantitatively by the so-called Markstein length (L_M) (Karpov et al., 1997). In terms of dimensionless numbers, the propagation speed of a stretched flame, S_L (relative to the unburned gas), can theoretically be related to that of an unstretched, planar flame, S_L^0, by the following relationship (this holds for conditions that are sufficiently far away from extinction, as the flame can extinguish at a high enough value of stretch):

$$\frac{S_L^0}{S_L} = 1 + \text{Ma}\,\text{Ka} \tag{2.6}$$

In Eq. (2.6), Ma and Ka are the Markstein and Karlovitz numbers, respectively, which are defined as follows:

$$\text{Ma} = \frac{L_M}{\delta_L} \tag{2.7}$$

$$\text{Ka} = \tau_C K = \frac{\delta_L K}{S_L} \tag{2.8}$$

For a given fuel, the Markstein number depends on several parameters including the equivalence ratio and the effective Lewis number, and can also take negative values. Markstein number reflects the stability of flame with respect to stretch. Positive values of Ma indicate a decrease of flame speed with the increase of stretch rate. In this case, if any outgrowth appears at the flame front (stretch increase), the flame speed in the corresponding position will be decreased, and this makes the flame stable. In contrast, a negative Ma means that the flame speed increases with the flame stretch rate. In this case, if any protuberance appears at the flame front, the flame speed in this position will be increased, leading to an unstable situation.

2.2 TURBULENT PREMIXED FLAMES

2.2.1 Structure of Turbulent Premixed Flames

In gas turbine combustors, like in many other practical combustion devices, turbulent flow conditions prevail. The degree to which the turbulent fluctuations affect the combustion and flame structure can be discussed on dimensional grounds, based on the relevant characteristic time and length scales. A short overview is provided here, whereas much more detailed discussion can be found in the literature (Libby and Williams, 1994). A representative root mean square of the velocity fluctuation shall be denoted by u', which can be associated with a turbulence kinetic energy k_0 ($k_0 = 0.5u'^2$). The length scale characterizing the large eddies shall be denoted l_0. Based on these scales, a turbulence Reynolds number can be defined, which can also be related to the scales of a laminar flame, assuming Pr = 1, as follows:

$$\mathrm{Re}_\mathrm{T} = \frac{u' l_0}{\nu} = \frac{u' l_0}{S_\mathrm{L} \delta_\mathrm{L}} \tag{2.9}$$

In the following a sufficiently large Re_T is assumed. A timescale associated with l_0 can be obtained as follows:

$$\tau_\mathrm{T} = \frac{l_0}{u'} \tag{2.10}$$

For high values of Re_T, based on the energy cascade view of turbulence, the dissipation rate ε of velocity fluctuations can be estimated as follows:

$$\varepsilon = \frac{u'^3}{l_0} \tag{2.11}$$

The turbulence energy is dissipated by viscous action at the lowest end of the cascade, by the smallest eddies. The associated characteristic length and timescales are the so-called Kolmogorov length (η) and time (τ_η) scales (Libby and Williams, 1994):

$$\eta = \left(\frac{\nu^3}{\varepsilon}\right)^{1/4}; \quad \tau_\eta = \left(\frac{\nu}{\varepsilon}\right)^{1/2}, \text{ with } \frac{\eta}{l_0} = \frac{1}{\mathrm{Re}_\mathrm{T}^{3/4}} \tag{2.12}$$

The flame surface is wrinkled and stretched by the turbulence eddies. The flame stretch can be related to the Kolmogorov timescale. In combination with the chemical timescale (τ_C), the following definition of the Karlovitz number can be obtained, which can also be related to the ratio of the flame thickness to Kolmogorov length scale, assuming Pr = 1:

$$\mathrm{Ka}_\mathrm{T} = \frac{\tau_\mathrm{C}}{\tau_\eta} = \left(\frac{\delta_\mathrm{L}}{\eta}\right)^2 \tag{2.13}$$

Comparing the turbulence timescale τ_T (Eq. 2.10) with the chemical timescale τ_C (Eq. 2.4), a Damköhler number can be defined as follows:

$$\mathrm{Da}_\mathrm{T} = \frac{\tau_\mathrm{T}}{\tau_\mathrm{C}} = \frac{l_0}{\delta_\mathrm{L}} \frac{S_\mathrm{L}}{u'} \tag{2.14}$$

Based on dimensional considerations, different regimes of premixed combustion can be identified in terms of the parameters presented earlier. These are presented graphically in the so-called Borghi diagram, sketched in Figure 2.3 (Borghi, 1988).

The region $\mathrm{Ka}_\mathrm{T} < 1$ ($\mathrm{Da}_\mathrm{T} > 1$), i.e., $\delta_\mathrm{L} < \eta$, defines the so-called "flamelet regime," which is bounded by the so-called Klimov–Williams line with $\mathrm{Ka}_\mathrm{T} = 1$, i.e., $\delta_\mathrm{L} = \eta$. Here, it is assumed that the smallest (Kolmogorov) eddies cannot penetrate into the flame zone and the role of turbulence remains solely restricted to the wrinkling and stretching the flame front that remains essentially laminar in its inner structure. Within the flamelet regime, two different regimes are identified. For $u'/S_\mathrm{L} < 1$, the flame front is solely wrinkled by turbulence (wrinkled flamelets). For $u'/S_\mathrm{L} > 1$, it is more strongly affected by turbulence and can exhibit folders entrapping burned or unburned gas (corrugated flamelets).

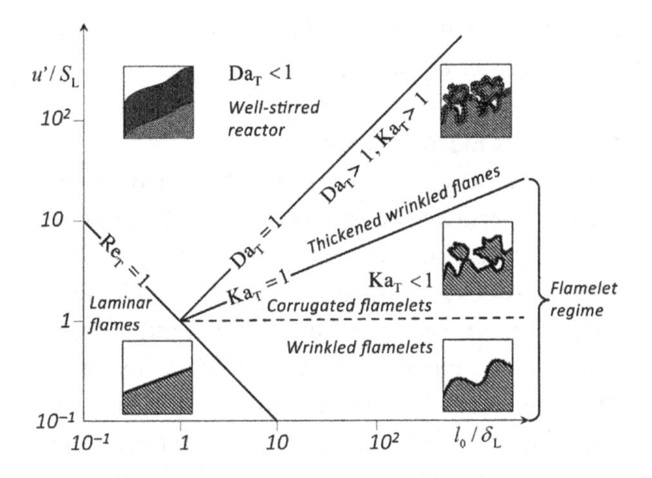

Fig. 2.3. Classification of regimes in premixed combustion after Borghi (1988).

Outside the flamelet regime, for $Ka_T > 1$, i.e., $\delta_L > \eta$, two regimes can be defined depending on the Damköhler number. For $Da_T > 1$ and $Ka_T > 1$, the flame structure is referred to as "thickened wrinkled flames" (or "distributed reaction zones"), where small eddies can penetrate into the flame zone and increase the heat and mass transfer. For $Da_T < 1$, the turbulent timescale is smaller than the chemical timescale. Even the largest eddies can access the thickened reaction zone and an intense mixing takes place through all scales. This zone is referred to as "well-stirred reactor," where no distinctive flame front exists in a strict sense.

The Borghi diagram results from theoretical considerations based on dimensional grounds and should be interpreted as such. The underlying combustion conception is idealized insofar that it assumes single-step chemistry, high activation energy, Le = 1, equal diffusivities of reactants, and steady and spatially uniform characteristics. The importance of unsteady effects was demonstrated by Poinsot et al. (1995), who showed that the smallest eddies cannot disturb the inner structure of a laminar flame, even though the flame thickness is of similar size, since the lifetime of the eddies is too short. Thus, for the existence of the flamelets, the condition of $Ka_T \leq 1$ of the Borghi diagram was found to be too restrictive, where flamelets were observed to exist up to $Ka_T = 16$ (Poinsot et al., 1995). It is also argued that increasing turbulence level favors the formation of even thinner highly strained reaction layers, so that the flamelets may extinguish before they broaden, and a thickening by the penetration of small eddies is more likely to occur in the preheat zone, but not in the reaction layer (Lipatnikov and Chomiak, 2002).

2.2.2 The Turbulent Flame Speed

While the laminar flame speed (S_L) is a well-defined property based on thermochemical properties of the mixture, the turbulent flame speed (S_T) is strongly affected by turbulence, and its determination is a main challenge in turbulent premixed combustion (Libby and Williams, 1994). In general, turbulence increases S_T to values much larger than S_L. The basic mechanism for the increased burning velocity by turbulence is the increase of the flame surface area (compared with the area of the approach flow) by the wrinkling action of the turbulent eddies, where the wrinkled flame front, the brush, occupies a finite space in the propagation direction, called the brush thickness. However, turbulence levels beyond a certain value cause only a small increase, if at all, and may lead to quenching of the flame (Law, 2006).

Turbulent flame speeds measured by Kobayashi et al. (1996) for methane–air flames, for $\phi = 0.9$, at different pressures show that S_T/S_L increases monotonically with increasing non-dimensional fluctuation velocity (u'/S_L), though with a gradually decreasing slope. This is known as the bending effect. A qualitative explanation to this behavior can be given as follows: as u' increases, the wrinkling of the flame surface increases. This, however, causes the brush to propagate faster, which makes the (e.g. curved) brush shorter. This, in turn, allows the brush less space over which the wrinkling can occur. These two competing effects may cause the bending of the curves (Driscoll, 2008). Another explanation for the bending of the curves can be given by the reduction in the local laminar flame speed with increasing strain and local quenching. Beyond a certain value of u', the surface density and brush thickness cannot grow indefinitely. Flamelets eventually merge and extinguish due to strain. Prior to a complete quenching of the flame, after having reached a maximum value, a mild decline of the S_T/S_L curve is usually observed (Abdel-Gayed et al., 1984) by a further increase of u'/S_L. Note that S_T/S_L increases with pressure (Kobayashi et al., 1996). This is mainly due to the decrease of S_L with pressure for methane–air flames, as already discussed earlier ($S_L \sim p^{-0.5}$), as the measured S_T (Kobayashi et al., 1996) remained practically uninfluenced by pressure. Expressions for calculating turbulent flame speed will be discussed in more detail in the following chapters.

2.3 SWIRL FLOW AERODYNAMICS

Swirling flows are frequently utilized in technical combustion systems for enhancing mixing and stabilizing combustion. In lean premixed technology, significantly high degrees of swirl are applied for stabilizing the flame and achieving a high power density and good burnout, under lean conditions. For theoretical analyses, swirl velocity profiles in combustion chambers are often approximated by the so-called Rankine vortex, which consists of a forced vortex (solid body rotation) in the core, and a surrounding free vortex (potential vortex) (Munson et al., 2009). Denoting the angular speed and radial extension of the vortex core by Ω and r_C, respectively, which results in a maximum swirl velocity of $W_{max} = \Omega r_C$ at the edge of the vortex core, the radial swirl velocity profile of a Rankine vortex can be expressed as follows:

$$w(r) = \begin{cases} W_{max} \dfrac{r}{r_C}, \text{ for } r \leq r_C \\ W_{max} \dfrac{r_C}{r}, \text{ for } r > r_C \end{cases} \tag{2.15}$$

The level of swirl in combustors is usually characterized by the swirl number (S), which is defined as the ratio of axial flow of tangential momentum to the axial flow of axial momentum given as follows (where contributions by the static pressure and turbulent fluctuations are neglected, as frequently done in applications):

$$S = \frac{\int_0^R \rho u w r^2 \, dr}{R \int_0^R \rho u^2 r \, dr} \tag{2.16}$$

where R is the combustor radius.

2.3.1 Vortex Breakdown

A unique characteristic of the swirling flows is the so-called vortex breakdown (VB) (Sarpkaya, 1971; Escudier, 1988; Keller, 1995), which can occur under certain flow conditions. VB is a sudden change of the vortex structure, which is usually marked by abrupt swelling and formation of a stagnation point on the axis of rotation with flow reversal downstream of it. The central recirculation zone behind the stagnation point is often referred to as the internal recirculation zone (IRZ), which enhances the mixing and combustion and acts as an aerodynamic flameholder. In combustion systems, the swirling flows almost always exhibit a VB, since this is a very effective, and, thus, favored, method of flame stabilization.

In a swirling flow, whether a VB occurs or not is primarily controlled by the swirl number. If the swirl number is high enough, i.e., if it is above a critical value ($S > S_{cr}$), VB occurs. However, it is not possible to assign a universal value to S_{cr} for all types of swirl flows (usual values varying in the range $\sim 0.4 < S_{cr} < \sim 1$), since S_{cr} depends on particular features of the flow, such as the velocity profile shapes. There are several theories for explaining the underlying mechanisms of VB. According to the "theory of critical state" (Squire, 1962; Benjamin, 1962), swirling flows can be categorized as supercritical ($S < S_{cr}$) flows, which do not allow an upstream propagation of inertia waves, and the subcritical ($S > S_{cr}$) flows, which allow an upstream inertia wave propagation. A quite frequently encountered VB mechanism in gas turbine combustors is defined as a transition from the supercritical state to the subcritical state (Keller, 1995). Downstream the VB, the supercritical state may be recovered, if the swirl number reduces below the critical value due to, e.g., increase of the axial velocity due to thermal expansion caused by combustion. VB can take place in different forms. Two basic forms are the "bubble type" and the "spiral type" (Lucca-Negro and O'Doherty, 2001).

The spiral type is usually observed for rather low swirl numbers. The bubble type, which is normally observed in gas turbine combustors, is likely to occur at high swirl numbers. In bubble-type VB, under certain conditions, the vortex core may extend far downstream behind the bubble and perform a spiraling motion (similar to the spiral-type VB), which is then referred to as the "precessing vortex core."

For an axisymmetric, slender Rankine vortex, Escudier (1988) developed the following expression for the axial pressure gradient on the axis:

$$\left.\frac{1}{2\rho}\frac{\partial p}{\partial x}\right|_{r=0} = \left.\frac{1}{2\rho}\frac{\partial p}{\partial x}\right|_{r\to\infty} + \frac{W_{\max}^2}{r_c}\frac{dr_c}{dx} \tag{2.17}$$

where it is expressed as the sum of two contributions, namely, an externally imposed axial gradient (the first term on the right-hand side) and a contribution by the vortex conditions in the core (the second term on the right-hand side). One can see that an expansion of the core ($dr_C/dx > 0$) contributes to generation of an adverse pressure gradient on the axis, and, thus, triggers VB and formation of a recirculation zone, This effect increases with decreasing core radius and increasing maximum swirl velocity on the core edge (Eq. (2.17)).

Brown and Lopez (1990) pointed out the role of azimuthal vorticity on VB. Assume an axisymmetric swirl flow, where the azimuthal and radial vorticity components are zero, at time $t = 0$. For that case, for the initial development of the azimuthal vorticity (ω_θ), the following equation can be deduced from the vorticity transport equation (Fritz, 2003):

$$\left.\frac{\partial \omega_\theta}{\partial t}\right|_{t=0} = \omega_x \frac{\partial w}{\partial x} \tag{2.18}$$

Thus, a decrease of the swirl velocity in the axial direction, e.g., due to an expansion of the vortex core in combination with angular momentum conservation, leads to generation of negative azimuthal vorticity, which can, according to the Biot–Savart law, induce negative axial velocities in the core flow, and, thus, trigger VB.

2.3.2 Swirl Burners

In lean premixed technology, under very lean conditions, the degree of applied swirl is especially high for enhancing the mixing and flame stabilization. A premix swirl burner with indication of some important flow elements and structures is qualitatively sketched in Figure 2.4.

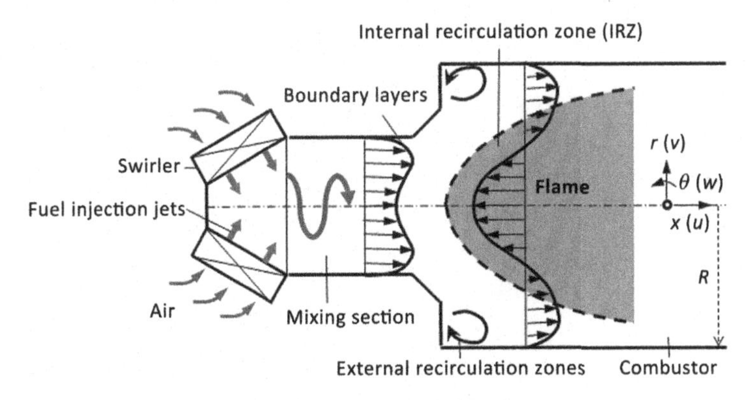

Fig. 2.4. Qualitative sketch of flow structures in a premix swirl burner: x, r, and θ, axial, radial, and azimuthal coordinates; u, v, and w: time-averaged axial, radial, and swirl (azimuthal) velocity components.

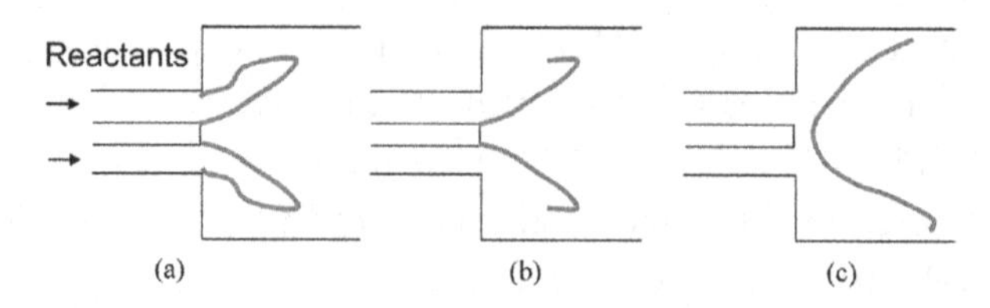

Fig. 2.5. Qualitative sketch of possible flame configurations in a premix swirl burner with centerbody (Lieuwen et al., 2008b).

The burner sketched in Figure 2.4 does not have a "centerbody." Many swirl burners are designed with a centerbody, which extends through the axis of the mixing section, forming an annular space for the swirling flow. In such configurations, the flow in the combustor exhibits a recirculation bubble in the wake of centerbody, which quite often merges with the recirculation zone created by the swirl flow, to form the IRZ. In this configuration, the flame is often stabilized at the centerbody, in the inner and outer shear layers, as sketched in Figure 2.5a. If the strain rate in the shear layers is too large, the flame locally extinguishes, and moves downstream. Figure 2.5b shows a sketch for this situation occurring in the outer shear layers. Figure 2.5c illustrates a situation where the flame strain is too large in the inner and outer shear layers and the flame is stabilized by the VB bubble farther downstream.

CHAPTER 3

Properties of Hydrogen-Containing Fuels

Hydrogen has a very wide flammability range; it is easy to ignite, and has a large flame propagation velocity and small quenching distance. Several combustion-relevant properties of hydrogen and methane are provided by Schefer et al. (2008). Hydrogen has a higher heating value relative to methane per unit mass. It is important to note, however, that per unit volume the lower heating value for H_2 is about a factor of 4 less than CH_4. Significantly larger storage capabilities are therefore required for gaseous H_2. In a stoichiometric mixture of hydrogen/air at ambient pressure and temperature, the quenching distance is 0.64 mm while that for methane is approximately 2.5 mm. At higher pressure and temperatures these values are further reduced. The flammability limits of hydrogen are much wider than those of methane. The limits, by volume, are from 4% [lower flammability limit (LFL)] to 75% [upper flammability limit (UFL)] for mixtures with air. These values compare with methane flammability limits of 5.5% (LFL) to 15% (UFL). Clearly, the UFL for H_2 is significantly greater compared with that for methane and other conventional hydrocarbon fuels.

The flame propagation velocity in a mixture of hydrogen and air is very large in comparison to that of methane and air. This is due primarily to the faster reaction rates of the H_2/O_2 system, since the flame speed is approximately proportional to the square root of the reaction rate (Glassman and Yetter, 2008). The large diffusion coefficient of H_2 also plays a role in the large flame speed because of the enhanced transport of radicals and heat ahead of the flame (preferential diffusion). At 25°C and 1 atm, the flame speed in a stoichiometric mixture of hydrogen and air is 1.85 m/s while for methane it is 0.40 m/s. A further result of the faster reaction rates in hydrogen/air flames is a much thinner flame front than for methane and other hydrocarbons.

Measured laminar flame speeds (S_L^0) of various fuel–air mixtures, including hydrogen–air for a range of equivalence ratios (Φ) at NTP conditions, were provided by Law and Kwon (2004).

Flashback Mechanisms in Lean Premixed Gas Turbine Combustion

The major constituents of syngas fuels are H_2, CO, and N_2. However, the performance of mixtures cannot always straightforwardly be deduced from the knowledge on the individual components, due to their interactions.

3.1 FLAMMABILITY LIMITS OF HYDROGEN FUEL BLENDS

Flammability limits of CH_4/H_2/air mixtures are investigated experimentally and computationally using different configurations such as a flame tube and a bomb, by Van den Schoor et al. (2008). Table 3.1 presents the measured LFL and UFL for different fuel compositions. One can see that addition of H_2 enlarges the flammability limits. This effect is much more pronounced for the UFL, corresponding to rich mixtures. In predicting the flammability limits of fuel mixtures, quite often, the so-called Le Chatelier's rule (Mashuga and Crowl, 2000) is employed, which is expressed as follows:

$$L_{\mathrm{m}} = \left(\sum_{i}^{n} \frac{X_i}{L_i} \right)^{-1} \tag{3.1}$$

where L_{m} is the calculated flammability limit of the fuel mixture, L_i is the flammability limit of the ith fuel component, and X_i is the volume fraction of the ith component in the mixture. The accuracy of the rule can be tested based on the data presented in Table 3.1. For the UFL, we have $L_{\mathrm{CH_4}} = 15.8\%$ (Table 3.1) and $L_{\mathrm{H_2}} = 75\%$ (Schefer et al., 2008). For $X_{\mathrm{H_2}} = 20\%$, 40%, and 60%, we obtain by the Le Chatelier's rule (3.1) the following values for the UFL of the mixture: L_{m} = 18.8%, 23.1%, and 30.0%. These values are quite close to the measured values in Table 3.1, with deviations about 1%, 5%, and 7%, respectively. This

Table 3.1. Experimentally Determined Flammability Limits of Methane/Hydrogen/Air Mixtures (Van den Schoor et al., 2008)

Fuel Composition		LFL		UFL	
CH_4 (mol%)	H_2 (mol%)	mol%	Φ	mol%	Φ
100	0	4.4	0.438	15.8	1.787
80	20	4.2	0.355	19.0	1.899
60	40	4.0	0.278	24.2	2.128
40	60	4.0	0.218	32.4	2.511

comparison indicates that the Le Chatelier's rule can be used to obtain reasonable guesses for the flammability limits of gaseous combustible mixtures, where the inaccuracy seems to increase with increasing hydrogen content.

3.2 LAMINAR FLAME SPEED OF HYDROGEN FUEL BLENDS

The laminar flame speed of a mixture does not vary linearly between the respective pure values of the mixture components, i.e., $S_{L,mix} \neq X_1 S_{L,1} + X_2 S_{L,2}$. Di Sarli and Di Benedetto (2007) calculated the laminar flame speeds of hydrogen–methane/air mixtures at NTP conditions using CHEMKIN PREMIX code with the GRI-Mech kinetic mechanism, for different equivalence ratios and fuel composition.

Figure 3.1 displays the variation of the (uncurved, unstretched) laminar flame speed as a function of the hydrogen content at three values of the equivalence ratio at NTP conditions. One can observe that the values of the blends of the laminar flame speeds are always smaller than those obtained by averaging those of the pure fuels according to their molar proportion. Linear trends can, however, be found for certain ranges of the fuel composition (Figure 3.1). It is possible to identify three regimes in the hybrid flame propagation depending on the hydrogen mole fraction in the fuel. At low hydrogen contents ($0 < X_{H_2} < 0.5$), a regime of methane-dominated combustion takes place, being characterized by a linear and slight increase of the methane laminar burning velocity on adding hydrogen. At high hydrogen contents ($0.9 < X_{H_2} < 1$), a regime of methane-inhibited hydrogen combustion occurs corresponding to a linear and sharp decrease of the hydrogen laminar burning velocity on increasing the methane presence in the fuel.

Expressions were developed to calculate the laminar flame speeds of H_2/CH_4 fuel blends depending on the laminar flame speeds of the components, the fuel composition, and the equivalence ratio (ϕ). Di Sarli and Di Benedetto (2007) proposed a Le Chatelier's rule–like expression, given as follows:

$$S_L(\phi, X_{H_2}) = \frac{1}{X_{H_2} / S_{L\_H_2}(\phi) + (1 - X_{H_2}) / S_{L\_CH_4}(\phi)} \quad (3.2)$$

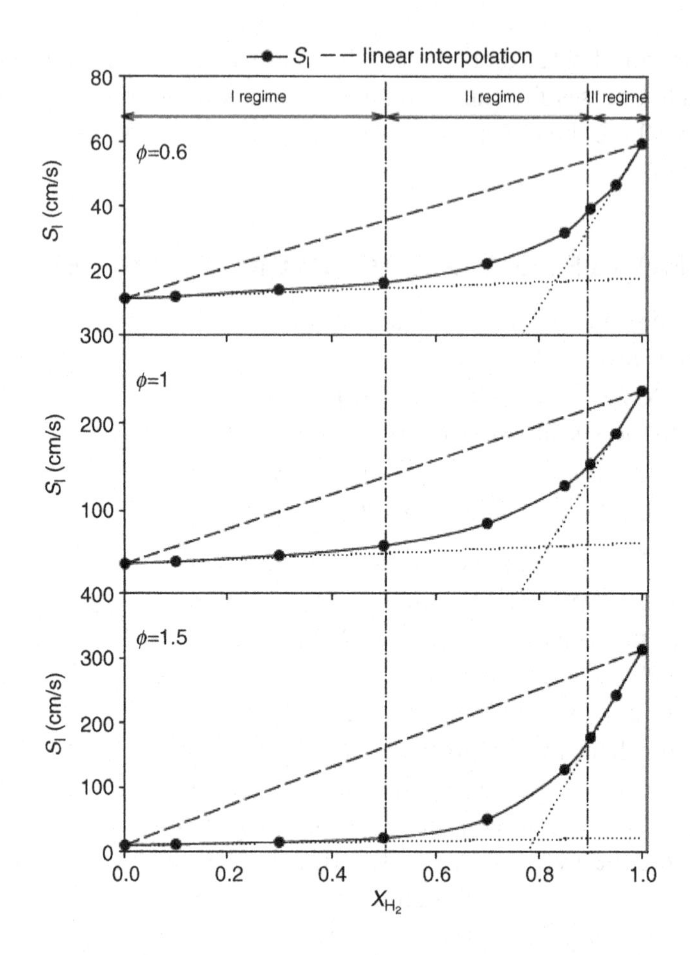

Fig. 3.1. Calculated laminar flame speed of hydrogen–methane/air mixtures as a function of the hydrogen content at three values of the equivalence ratio, at NTP conditions (Di Sarli and Di Benedetto, 2007).

where $S_{L\_H_2}$ and $S_{L\_CH_4}$ are the laminar flame speeds of hydrogen and methane. Mole fraction of hydrogen in the fuel is denoted by X_{H_2}.

Figure 3.2 shows variation of the laminar flame speed with equivalence ratio, for different values of hydrogen content in the fuel ($X_{H_2} = 0.1$, 0.5, 0.7, 0.85, and 0.95), as calculated by CHEMKIN and obtained via Eq. (3.2) at NTP conditions. It appears that at lean and stoichiometric conditions a good agreement is obtained. It is interesting to note that this applies also to the rich mixtures with rather low hydrogen content ($X_{H_2} < 0.7$).

Huang et al. (2006) presented an experimental work on the laminar flame speed of natural gas–hydrogen–air mixtures in a constant-volume

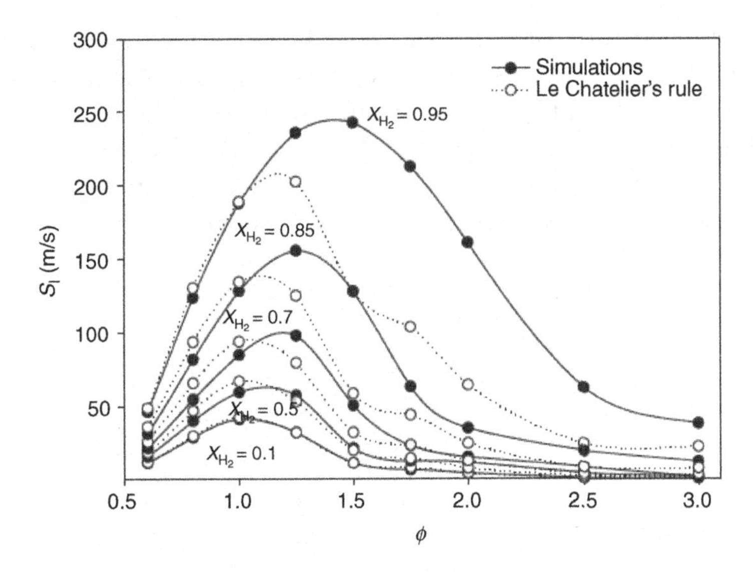

Fig. 3.2. Calculated and correlated variation of the laminar flame speed as a function of the equivalence ratio for different values of the hydrogen content in the hydrogen–methane/air mixtures, at NTP conditions (Di Sarli and Di Benedetto, 2007).

bomb at NTP conditions, where flame stretch effects were also investigated. The measurements showed that the Markstein length (and Markstein number) decreases with the increase of hydrogen fraction. This implies a decrease of the flame stability with increasing hydrogen content.

Insight into the flame speed characteristics of CO/H_2 syngas mixtures with varying degrees of methane cofiring can be acquired from the results of Lieuwen et al. (2008a), which are shown in Figure 3.3, as the dependence of the flame speed (uncurved, unstretched) on fuel

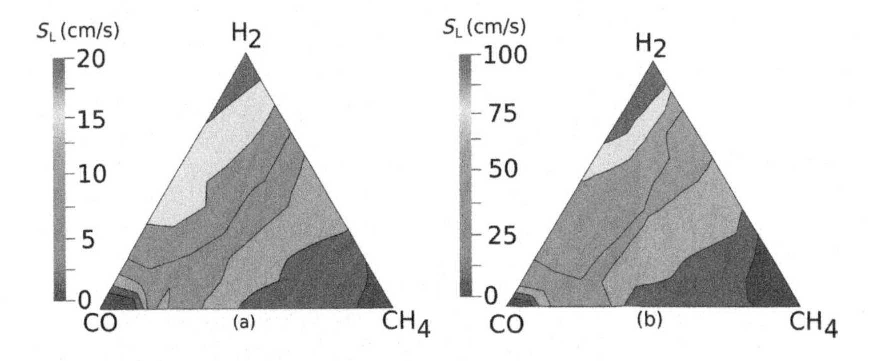

Fig. 3.3. Laminar flame speed (cm/s) as a function of fuel composition at fixed 1500 K (a) and 1900 K (b) adiabatic flame temperature at 4.4 atm with 460 K reactant temperature (Lieuwen et al., 2008a).

composition at two fixed adiabatic flame temperatures, namely, at 1500 and 1900 K (at 4.4 atm, with 460 K reactant temperature). The volumetric fuel composition is given by the location within the triangle, where the three vertices denote pure CO, H_2, or CH_4. At each point, the mixture equivalence ratio was adjusted such that the mixture has the given flame temperature. As expected, the high H_2 mixtures have the largest flame speeds.

CHAPTER 4

An Overview of Flashback Mechanisms

Flame flashback can be defined as an upstream propagation of flame back into the burner. This can be caused by an imbalance in the local flame and flow speeds, when the former exceeds the latter. This is the mechanism that is normally associated with the concept of flashback. However, some authors have rather generalized the definition of flashback to cover all mechanisms that cause undesirable combustion in the upstream part of the burner. For example, autoignition was considered as a flashback mechanism in some rather earlier sources (Lefebvre, 1983), since its consequence is practically the same as an upstream propagating flame, i.e., an undesirable combustion in the burner. On the other hand, in rather recent studies (Fritz et al., 2004), autoignition is not considered as a form of flashback, since the mechanism through which it occurs is not related to flame propagation. In order to be comprehensive, autoignition is included in the present essay, as its consequences are similar to flame propagation flashback mechanisms.

For a swirl burner (which is commonly encountered in industrial gas turbines), there are several modes of flashback, leading to the following mechanisms that will be addressed in the present contribution and will be discussed in depth in later chapters:

1. Autoignition: This may occur if the autoignition delay time is less than the residence time within the burner. In practical systems this may occur due to a reduction in flow velocity or to an increase in temperature of the fuel/air mixture. The fuel/air mixture temperature may increase due to convective heating from the surfaces of the burner, which may be heated by radiative heat feedback from the combustor. Additionally the autoignition delay time may reduce due to fuel composition changes.
2. Flame propagation:
    a. Combustion instabilities: This may occur if the combustion process induces large pressure fluctuations within the combustor. Sustained operation with high levels of pressure fluctuations need to be avoided, as the combustor system may sustain damage due

to fatigue. However, transient events of high-pressure fluctuations are commonplace. Such transient events lead to very low burner flow velocities, such that the flame propagates deep into the burner, due to core flow and/or wall boundary layer flashback. Burner damage however takes a finite time to occur, which is typically much longer than such transient events. Damage can therefore be avoided if, after such an event, the flame moves back into the combustor and does not remain in the burner. Although propagation is the mechanism that leads to flashback in this case, the avoidance of combustor damage relies on flame quenching and extinction.

b. Flame propagation in the core flow: This may occur if there is an increase in the turbulent burning velocity or a decrease in the flow velocity. The former may result if there is an increase in the flame temperature or fuel reactivity due to fuel composition changes. Within a multiburner arrangement, depending on the burner characteristics, flame propagation into one of the burners may result in an increased pressure drop coefficient for that burner. The air flow rate to this burner, and, there fore the velocity decreases,which further enhances the flashback event.
c. Flame propagation within boundary layers: This may occur due to similar reasons as for core flow propagation. However, in this case wall temperature also plays a key role.
d. Combustion-induced vortex breakdown (CIVB): The mechanisms listed earlier apply to any premixed burner; however, CIVB is specific to swirl-stabilized burners. In this case, the combustion process alters the burner fluid dynamics, such that the vortex breakdown bubble is moved from the burner exit region to deep within the burner. One of the key mechanisms for this is misalignment between surfaces of constant pressure, generated by the swirling flow, and surfaces of constant density, generated by the combustion, which leads to a baroclinic torque that encourages negative velocity along the burner axis.

CHAPTER 5

Flashback by Autoignition

Autoignition is the spontaneous ignition of a combustible mixture due to the thermodynamic state of the system, in which reactions are initiated with liberation of sufficient heat to start and sustain combustion, without the need of an external ignition source. In the literature (Glassmann and Yetter, 2008), there are tabulated values for the autoignition temperature (at atmospheric pressure) of mixtures of different fuels with air, which can be used as a first orientation by the combustion engineer.

Normally, not single values, but ranges are given for the autoignition temperature, which can sometimes be quite large, e.g., for hydrogen. These tabulated values refer to the temperatures above which the mixture will spontaneously ignite, if given sufficient time.

The onset of autoignition requires a certain amount of time, which is called (auto) ignition delay time (τ_{ig}). Depending on the initial thermodynamic state of the mixture, the autoignition time can vary between hundreds of microseconds and several seconds.

In premixed gas turbine combustion, long residence times in the premixer are beneficial in order to achieve a homogeneous mixture. However, this perspective is in contradiction with the requirement of avoiding autoignition. It is obvious that the residence time needs to be shorter than the autoignition delay time for a safe operation, but still long enough to achieve a good mixing. A safe and efficient design should, thus, provide an optimal balance of both effects, which requires knowledge on timescales relating to mixing and autoignition. It should be noted that the complex flow and mixing patterns that normally occur in practical combustors, rich in inhomogeneities and gradients, make it difficult to assign a single timescale for the autoignition delay time. For example, even if a very small portion of combustible mixture at a certain local mixing ratio experiences a long enough residence time that is sufficient for autoignition to occur, its

Flashback Mechanisms in Lean Premixed Gas Turbine Combustion

autoignition can lead to the ignition of the complete field of combustible gas in the premixer.

Information on the ignition delay times can be attempted to be obtained experimentally or computationally. As far as the experimental data are concerned, although there is a considerable amount of measured data on ignition delay times, the number of experimental investigations on hydrogen and hydrogen-containing fuels under gas turbine combustion–relevant conditions is rather limited. As far as the chemical kinetics calculations are concerned, the GRI-Mech 3.0 detailed mechanism, which is optimized for methane and natural gas, and universally used in kinetic calculations concerning such fuels, is quite often found not to provide sufficient accuracy for hydrogen and hydrogen/carbon monoxide mixtures. Thus, development of more accurate mechanisms is a current research field and some suggested mechanisms will be referred to in the following sections.

Lieuwen et al. (2008a) calculated ignition delay times for CH_4/H_2 and CO/H_2 blend fuels using detailed chemical reaction mechanisms. For CH_4/H_2 blend fuels, the GRI-Mech 3.0 reaction mechanism was used. For the CO/H_2 blend fuels the reaction mechanism of Davis et al. (2005) was employed, which is an optimized mechanism for such fuels. Figure 5.1 shows the variation of the ignition delay time (Lieuwen et al., 2008a)

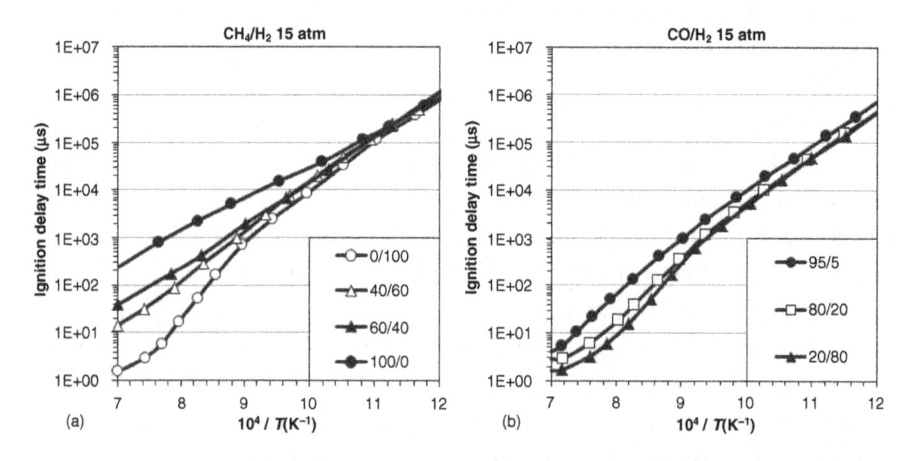

Fig. 5.1. Ignition delay time as a function of temperature for different fuel compositions for (a) CH_4/H_2 and (b) CO/H_2 mixtures at 15 atm and $\phi = 0.4$.

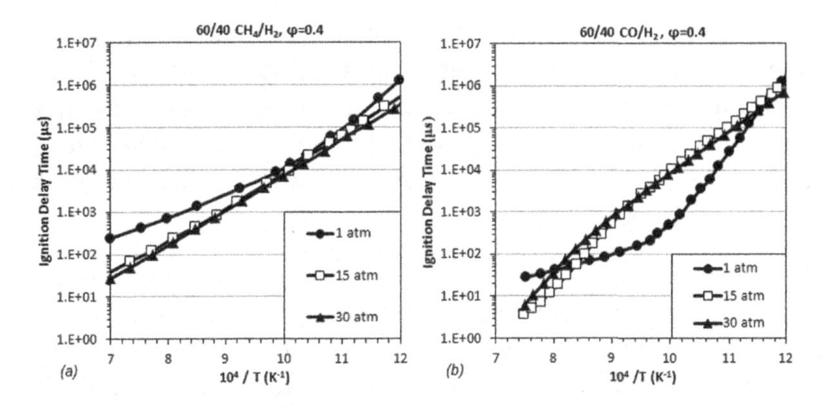

Fig. 5.2. Ignition delay time as a function of temperature at different pressures for 60/40 blend of (a) CH_4/H_2 and (b) CO/H_2 for $\phi = 0.4$.

with the unburned mixture temperature, for different compositions of CH_4/H_2 and CO/H_2 mixtures. At 15 atm, for $\phi = 0.4$, the corresponding adiabatic flame temperatures are around 1500 K.

As one would expect, the ignition delay time decreases with increasing H_2 content (Figure 5.1). It is interesting to note that this ignition-enhancing effect of hydrogen is most pronounced at rather high temperatures, i.e., $T > 1000$ K. In the low-temperature region ($T < 1000$ K), the ignition delay times do not considerably differ. Since the mixture temperature in the premixers of nonrecuperated gas turbines is normally lower than 1000 K, this prediction is encouraging, since it suggests the ignition delay times will not remarkably change by hydrogen addition, under these conditions. The effect of pressure is illustrated in Figure 5.2 where the predicted (Lieuwen et al., 2008a) ignition delay times as a function of temperature are displayed for different pressures ranging from 1 to 30 atm, for 60/40 CH_4/H_2 and CO/H_2 blends at $\phi = 0.4$.

Although one might expect a decrease of ignition delay time with increasing pressure, the results show that it is not always the case. For example, for the investigated CH_4/H_2 blend (Figure 5.2a), for 1000 K, the ignition delay time practically does not change with pressure. The shape of the 1 atm curve reveals the transition from the first to second explosion limit. The mixed effect of pressure is even more pronounced for the investigated CO/H_2 mixture (Figure 5.2b). For temperatures between about 1200 and 900 K, the 1 atm ignition delay times are predicted to be shorter compared with 15 and 30 atm cases. At higher temperatures, the predicted ignition delay times for 15 atm are shorter than those of the 30 atm ones.

Mittal et al. (2006) studied autoignition of H_2/O_2 and $H_2/CO/O_2$ mixtures in a rapid compression machine experimentally, and computationally. In addition to the GRI-Mech 3.0, the mechanism of Davis et al. (2005) and the mechanism of Li et al. (cited in Mittal et al., 2006) were also used. Experimental and calculated ignition delay times as a function of the CO content ($R_{CO} = X_{CO} / (X_{H_2} + X_{CO})$) for an equivalence ratio of unity are plotted in Figure 5.3a (Mittal et al., 2006), for compression pressure and temperature of 30 bar and 1010.5 K. Experiments show an increase in the ignition delay time with increasing CO content. The mechanisms of Davies et al. and Li et al. show a similar performance, but do not agree well with measurements, as the predicted ignition delay time remains practically constant for $R_{CO} < 0.8$. Predictions using GRI-Mech 3.0 show an opposite trend of reduction in ignition delay with CO addition for $R_{CO} < 0.8$. The effect of equivalence ratio on ignition delay time is illustrated in Figure 5.3b (Mittal et al., 2006) for a compression pressure of 50 bar and compression temperatures 990 and 1009 K. One can see that the ignition delay time

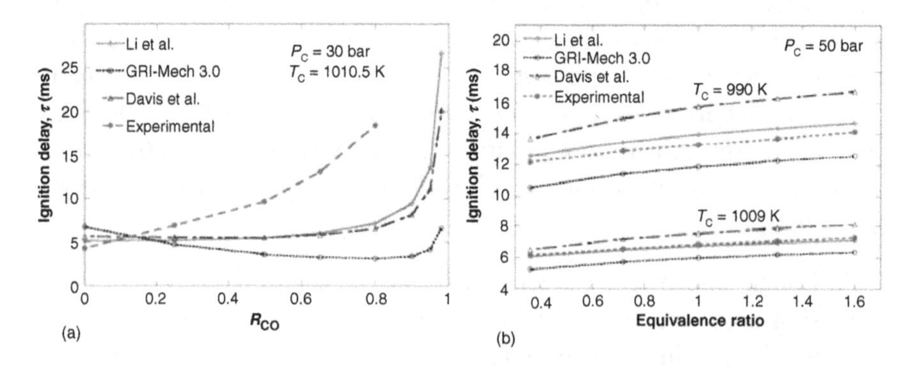

Fig. 5.3. Measured and predicted ignition delay times for a H_2/CO mixture. (a) Effect of CO addition. Molar composition: $(H_2 + CO)/O_2/N_2/Ar = 12.5/6.25/18.125/63.125$, $\phi = 1.0$. (b) Effect of equivalence ratio ($R_{CO} = 0.25$, $X_{H_2} = 6.667$, $X_{CO} = 2.222$) (Mittal et al., 2006).

increases only mildly with increasing equivalence ratio as observed in the experiments and qualitatively predicted by all mechanisms. Quantitative discrepancies between the experiments and the predictions, and between the reaction mechanisms, themselves, can, again, be observed. The authors (Mittal et al., 2006) discovered that the reaction $CO + HO_2 = CO_2 + OH$ is primarily responsible for the inaccuracy and obtained a much better agreement with the experiments for the considered cases, based on the reaction rate given by the mechanism of Li et al., by reducing the rate constant of this reaction by a factor of 4.

Beerer and McDonnel (2008) performed experimental and computational investigations on autoignition in hydrogen/air mixtures in a flow reactor at temperatures relevant to gas turbine combustion. The investigated (Beerer and McDonnel, 2008) continuous flow reactor, which was also taken as a basis in a previous study of Peschke and Spadaccini (cited in Beerer and McDonnel, 2008), can be considered to represent the conditions in the premixer of a gas turbine combustor more closely, compared with the other apparatus such as shock tube or compression machine. The computational investigations (Beerer and McDonnel, 2008) were based on homogeneous, zero-dimensional, isobaric, and adiabatic PFR simulations within the CHEMKIN code using the detailed reaction mechanism of O'Conaire et al. (2004). The calculated ignition delay times as a function of temperature were compared with their own measurements and measurements of other authors, for lean mixtures ($\phi = 0.2$–0.5) at pressures ranging from 1 to 23 atm, and shown in Figure 5.4.

As can be seen in Figure 5.4, the agreement of predictions with the measurements is fairly good, for temperatures above 1000 K. However, for lower temperatures ($T < 1000$ K), which are rather relevant for the premixers of gas turbine combustors, the simulated results do not agree at all with the measurements, where the measured ignition delay times are orders of magnitude shorter than the predicted values.

A similar and interesting comparison for hydrogen/carbon monoxide blends was provided by Lieuwen et al. (2008b), where the flow reactor data of Peschke and Spadaccini, Beerer et al., and Boleda et al. (all cited in Lieuwen et al., 2008b) were compared with the PFR predictions at different pressures using the reaction mechanism by Mueller et al.

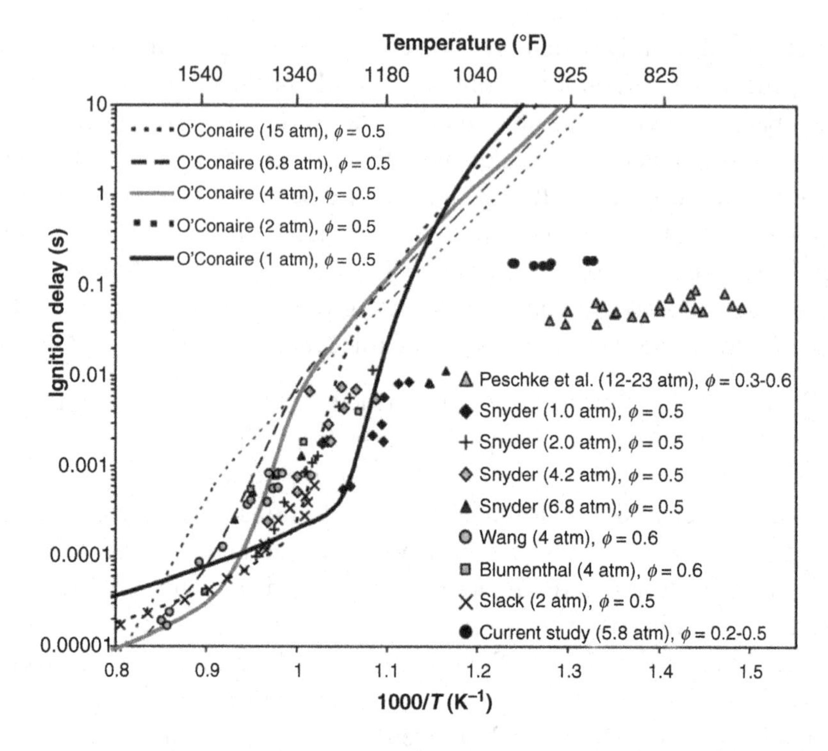

Fig. 5.4. Comparison of measured and calculated ignition delay times for hydrogen/air mixtures (Beerer and McDonnel, 2008).

(1999a, 1999b), as shown in Figure 5.5. Here, the different results obtained at different pressures were "corrected," assuming a pressure scaling by $p^{-0.75}$ as previously suggested by Peschke and Spadaccini (cited in Lieuwen et al., 2008b).

As one can see in Figure 5.5, different experimental results lie quite close to each other after pressure scaling. The predictions show deviations for high-temperature regions, which are not necessarily important for gas turbine applications. In the low-temperature region $T < 900$ K ($1000/T > 1.1$), relevant for gas turbine premixer conditions, predictions for different pressures are much closer, where, by trend, smaller ignition delay times are predicted for higher pressures. Nevertheless, as it was also observed in the comparison depicted in Figure 5.4, the predicted ignition delay times are one to two orders of magnitudes higher than the measured ones in the low-temperature region relevant for gas turbine applications (Figure 5.5).

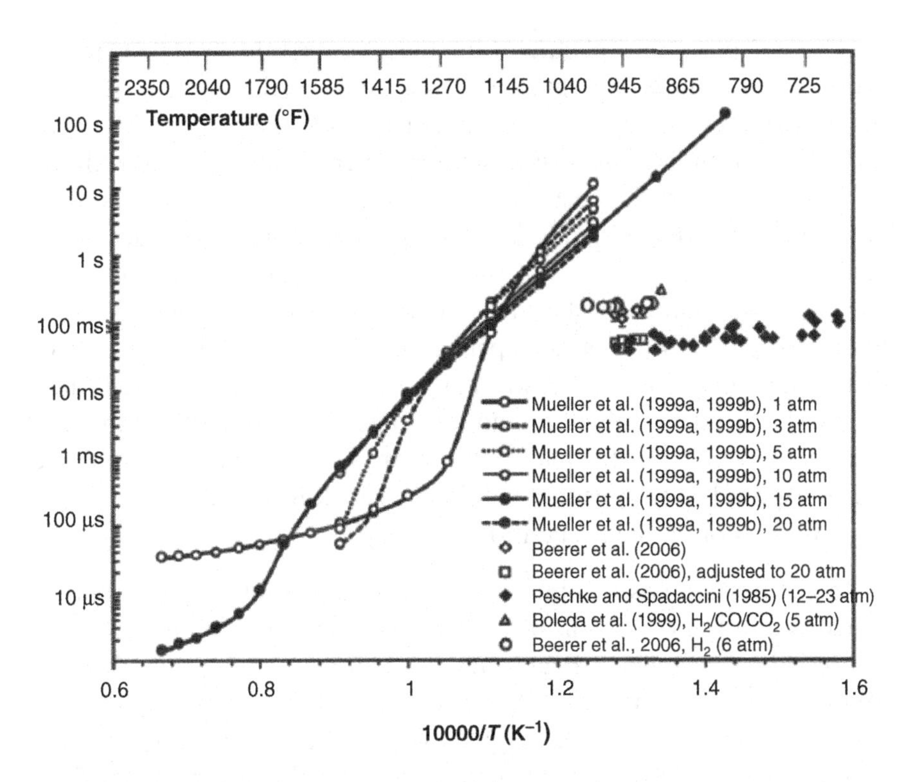

Fig. 5.5. Comparison of measured and calculated ignition delay times for hydrogen/carbon monoxide blends: 50% H_2 and 50% CO, after pressure scaling by $p^{-0.75}$ *(Lieuwen et al., 2008b).*

An analysis of the data points on the hydrogen–oxygen explosion limit chart revealed (Beerer and McDonnel, 2008) that the points exhibiting a discrepancy between the measurements and simulations (Figure 5.4) lie in the so-called "mild" ignition region (Yetter et al., 1991), which is bounded by the third limit and the "extended" second limit, on crossing over the extended second limit. Since in this region the reactions occur through the less reactive HO_2 and H_2O_2 radicals, an inaccuracy in the reactions involving these species was seen as a possible cause for the discrepancy. However, the analysis of Sabia et al. (2006) showed that the modifications to the mechanism required to have an agreement with the measurements lie well outside the established uncertainty factors attributed to these reactions, which ruled out the inaccuracies in the modeling of these reactions as the possible cause of the discrepancy. As a further possible cause of the observed discrepancy, influence of contaminants, particulate, and possible surface chemistry was mentioned (Lieuwen et al., 2008b).

Another possible explanation for the observed discrepancies in the predicted and measured ignition delay times in the low-temperature region (Figures 5.4 and 5.5) was postulated by Ströhle and Myhvold (2007). According to this reasoning, the experimental conditions in the mild combustion regime deviate so far from the assumed ideal homogeneous behavior that a simple homogeneous reactor alone is no longer sufficient to accurately predict the ignition delay time. This view was seen to be supported by the shock tube investigations (Voevodski and Soloukhin, 1965; Meyer and Oppenheim, 1971; Blumenthal et al., 1996) of ignition delay time. In the strong ignition region, the mixture heated by the shock tube resulted in a uniform ignition through the cross-section of the vessel. In the mild ignition, on the other hand, small flame kernels sporadically appeared within the shock-heated mixture, implying a very inhomogeneous ignition.

In the normal gas turbine operation, it is possible that nonhomogeneous ignition could still be present and therefore dominate the ignition behavior. Indeed, Beerer and McDonnel (2008) argue that the inhomegeneous ignition, which obviously cannot adequately be represented by a PFR, can be responsible for the observed discrepancy between the measured and calculated ignition delay times (Figure 5.4). Thus, they (Beerer and McDonnel, 2008) recommend the use of measurements and experimentally derived correlations for estimating the ignition delay times with a fair amount of accuracy. On the other hand, there are also differences in the experimentally obtained ignition delay times, depending on the apparatus used (constant-volume bomb, shock tube, rapid compression machine, flow reactors with different designs). Data obtained on flow reactors can be expected to represent the gas turbine premixer conditions at best. Still, it is obvious that one should, in any case, work with sufficiently large safety margins, due to the many uncertainties, especially for hydrogen and hydrogen blend fuels.

The following correlation for methane autoignition was suggested by Tsuobi (1975) based on shock tube measurements over a temperature range of 1200–2100 K, equivalence ratios between 0.5 and 2, and for total concentrations extending from 2×10^{-5} to 2×10^{-3} mol/cm^3, corresponding to pressures from 3 to 200 atm at 1800 K (ignition time in seconds, concentrations in mol/cm^3):

$$\tau_{ig} = 2.5\times10^{-15}\exp\left(\frac{26,700}{T}\right)[CH_4]^{0.25}[O_2]^{-1.02} \tag{5.1}$$

As can be deduced from Eq. (5.1), the ignition delay decreases with increasing temperature and pressure (through the concentrations).

Based on shock tube experiments, Petersen et al. (1996) suggested the following correlation for methane autoignition for a temperature range of 1400–2050 K, equivalence ratio range of 0.5–2.0, and for $[CH_4]$ and $[O_2]$ concentrations up to 3.6×10^{-5} mol/cm^3, for a wide pressure range, with an overall pressure dependence of $p^{-0.72}$:

$$\tau_{ig} = 4.05\times10^{-15}\exp\left(\frac{26,067}{T}\right)[CH_4]^{0.33}[O_2]^{-1.05} \tag{5.2}$$

Spadaccini and Colket (1994) performed ignition delay time measurements for methane and mixtures of methane with ethane, propane, and butane in a shock tube at pressures between 3 and 15 atm, temperatures from 1350 to 2000 K, and equivalence rations between 0.45 and 1.25. The following correlation was suggested (Spadaccini and Colket, 1994) for methane ignition delay time:

$$\tau_{ig} = 2.21\times10^{-14}\exp\left(\frac{22,659}{T}\right)[CH_4]^{0.33}[O_2]^{-1.05} \tag{5.3}$$

For mixtures of methane with ethane, propane, and butane in quantities up to 10%, the following correlation was suggested (Spadaccini and Colket, 1994) for temperatures between 1250 and 1880 K:

$$\tau_{ig} = 1.77\times10^{-14}\exp\left(\frac{18,693}{T}\right)[CH_4]^{0.66}[HC]^{-0.39}[O_2]^{-1.05} \tag{5.4}$$

In the above equation, the concentrations of ethane, propane, and butane are combined into a single hydrocarbon term [HC]. It was observed that even small amounts of higher hydrocarbon addition affect the chemical pathways, causing a reduction in the activation energy and the ignition delay time.

Li and Williams (2002) developed expressions for methane ignition delay time, based on a detailed chemical kinetics analysis. They developed two different expressions for high- and low-temperature ranges, while the dependence on methane concentration was removed by assuming the non-dimensional ignition time to be constant within the framework of a Frank-Kamenetskii approximation. The high-temperature expression, which was developed for temperatures above 1300 K (1300 K < T < 2000 K), is given as follows:

$$\tau_{ig} = 6.25 \times 10^{-16} \exp\left(\frac{23,000}{T}\right)[O_2]^{-1} \tag{5.5}$$

The above expression was observed to provide a very good agreement with the data of Spadaccini and Colket (1994). The proposed low-temperature expression limited to temperatures below 1300 K is provided as follows:

$$\tau_{ig} = 2.9 \times 10^{-13} T^{0.31} \exp\left(\frac{12,000}{T}\right)[O_2]^{-1} \tag{5.6}$$

Predictions by this expression were compared with the measurements of Petersen et al. (1999), which were obtained in a shock tube at elevated pressures of 40–260 bar, at three equivalence ratios (ϕ = 0.4, 3.0, 6.0), and within a temperature range from 1040 to 1500 K, and a very good agreement was observed (Li and Williams, 2002).

Calculated methane ignition delay times by the earlier correlations (5.1, 5.3, 5.5, 5.6), for $\phi = 0.5$ and $p = 1$ atm, are displayed in Figure 5.6 as functions of temperature (note that the correlations are displayed only within their declared temperature range of validity).

Measured shock tube data by different authors (Cowell and Lefebvre, 1986; Goy et al., 2001; Huang et al., 2004; DeVries and Petersen, 2007) are also indicated in the figure, as well as those of Spadaccini and Colket and Wagner (cited in Beerer, 2009).

In the high-temperature region (T > 1300 K), all measured data as well as the correlations of Spadaccini and Colket (5.3) and of Li and Williams (5.5) lie very close to each other. The correlation by Tsuobi (5.1) over-predicts the above-mentioned data and correlations throughout. In the low-temperature region (T < 1300 K), which is relevant for gas turbine premixers, the measured data show a large spread. The measurements

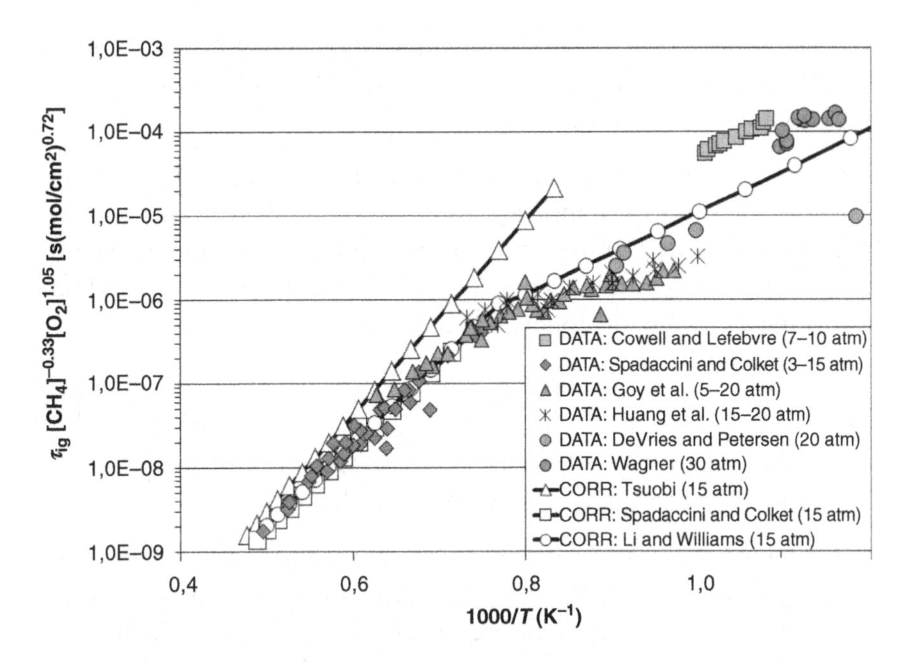

Fig. 5.6. Predicted methane ignition delay times by the correlations (CORR) for $\phi = 0.5$ and p = 15 atm as functions of temperature with indications of various shock tube measurements (DATA) for different equivalence ratios and pressures.

of De Vries and Petersen (2007), which are rather well represented by the low-temperature correlation of Li and Williams (5.6), lie more or less in the middle of the spread range (Figure 5.6).

A correlation for hydrogen autoignition delay time was provided by Peschke and Spadaccini (cited in Beerer and McDonnel, 2008), which was based on experimental data obtained on a continuous flow reactor. The experiments were performed for syngas ($H_2/CO/CO_2$). It was observed that the ignition delay time is nearly independent of the CO and CO_2 concentration and the autoignition characteristics of syngas are dominated by hydrogen kinetics. Thus, it was concluded that the measured syngas ignition delay results are meaningful for hydrogen/air autoignition. Based on this, for temperatures from 630 to 780 K, the pressure range from 12.3 to 23 atm, and equivalence ratios between 0.3 and 0.6, the following correlation for hydrogen autoignition results:

$$\tau_{ig} = 1.29 \times 10^{-7} \exp\left(\frac{3985}{T}\right)[H_2]^{-0.25}[O_2]^{-0.5} \tag{5.7}$$

This correlation represents the hydrogen low-temperature ($T < 800$ K) data of Peschke and Spadaccini displayed in Figure 5.4 very well, since it was based on this data set.

Cheng and Oppenheim (1984) performed experiments for methane/hydrogen mixtures for temperatures from 800 to 2000 K and pressures from 0.1 to 0.3 MPa and correlated the ignition delay time of the mixture ($\tau_{ig,M}$) to the ignition delay times of pure methane (τ_{ig,CH_4}) and pure hydrogen (τ_{ig,H_2}), by the following equation:

$$\tau_{ig,M} = \tau_{ig,CH_4}^{X_{CH_4}} \cdot \tau_{ig,H_2}^{X_{H_2}} \tag{5.8}$$

where X_{CH_4} and X_{H_2} denote the mole fractions of methane and hydrogen in the methane/hydrogen blend fuel.

The temperatures in the premixers of gas turbines are typically within the range from 600 to 900 K. The calculated autoignition times for methane following the low-temperature correlation of Li and Williams (5.6), and for hydrogen following the correlation of Peschke and Spadaccini

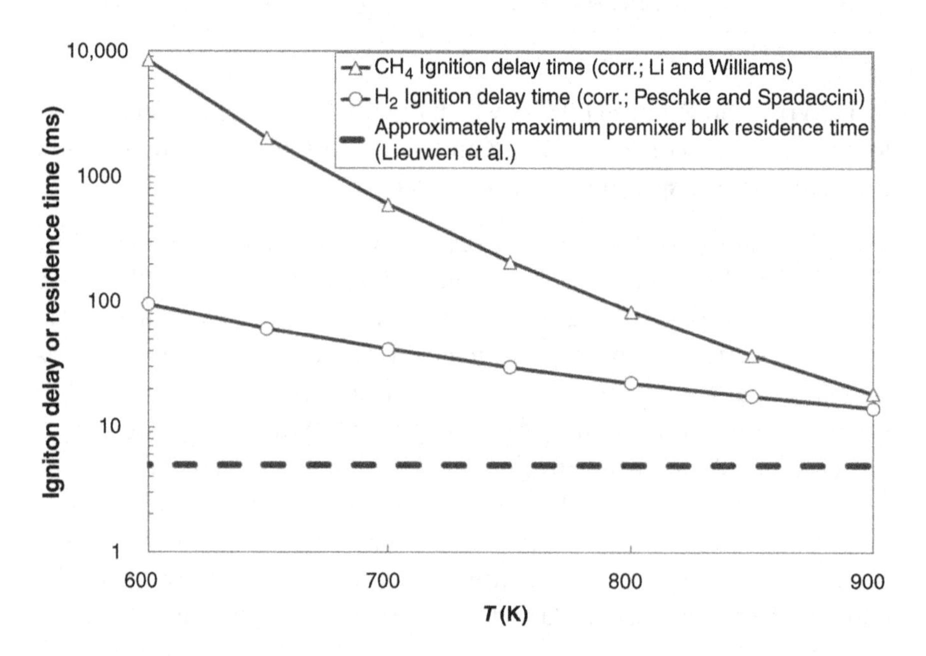

Fig. 5.7. Predicted autoignition delay times for methane (correlation: Li and Williams) and hydrogen (correlation: Peschke and Spadaccini) for $\phi = 0.5$ and p = 30 bar compared with approximate maximum premixer bulk residence time estimated by Lieuwen et al. (2008a).

(5.7), within this temperature range, for $\phi = 0.5$ and $p = 30$ bar, are displayed in Figure 5.7. The estimated approximate maximum gas turbine premixer residence time of 5 ms by Lieuwen et al. (2008a) is also marked in the figure. One can see that the ignition delay times predicted by the correlations are larger than the estimated maximum premixer bulk residence time. However, as already discussed, there are large uncertainties in predicting ignition delay times, especially for low temperatures. On the other hand, the flow inhomogeneities may lead to residence times in critical regions that are longer compared with the bulk residence time. Additionally radiative heat feedback from the combustor can lead to higher local temperatures close to the premixes walls. Therefore, for excluding the possibility of autoignition in the premixer, one should work with rather large safety factors.

CHAPTER 6

Flashback Due to Combustion Instabilities

Flashback can be triggered by velocity fluctuations in the burner associated with combustion instabilities. Combustion instabilities are characterized by large-amplitude pressure pulsations driven by unsteady heat release (Sattelmayer, 2003; Lieuwen et al., 2008a). A necessary condition for instability to occur is that the pressure and heat release oscillations are in phase. Changes in the fuel composition affect combustion instabilities by altering the phase angle. For combustion instability, two mechanisms are known to be especially important in premixed systems: the fuel/air ratio oscillations and the vortex shedding. In the former, acoustic oscillations in the premixer cause fluctuations in the fuel and/or air supply rates, leading to a periodically changing equivalence ratio. This is convected into the flame, where it creates heat release fluctuations that drive the instability. The coupling of the premixer acoustics with the fuel system is affected by the pressure drop across the fuel injector. The vortex shedding mechanism is due to large-scale, coherent vortical structures. These structures can be formed in association with flow separation from flameholders and rapid expansions, as well as vortex breakdown (precessing vortex core) in swirling flows. They distort the flame front and can cause oscillations in the heat release rate.

At high pulsation levels, significant periodic undershoots of the burner flow velocity below its time-averaged value can result. If the frequency is low enough an upstream propagation of the flame may occur, which can take place in the boundary layer, or in the core flow. In regular operation, high pulsation levels must be avoided for further reasons. Thus, flashback due to combustion instabilities should not occur during stable combustor operation, and this mechanism gains importance in the case of unexpected combustion instabilities.

Coats (1980) pointed out that in most cases flashback was established as an outcome of low-frequency flow instabilities. Keller et al. (1982)

Flashback Mechanisms in Lean Premixed Gas Turbine Combustion

performed experiments in a duct-shape combustor, where the flame was stabilized by a backward-facing step. They found that flashback occurs primarily as a consequence of a flow reversal triggered by large-scale vortex dynamics. The interaction between the recirculating vortices at the base of the step and the trailing vortices behind its edge was identified as the unstable process leading to flashback. The mechanism observed by Keller et al. (1982) is qualitatively illustrated in Figure 6.1. The top drawing shows the trailing vortex patterns for stable combustion, where two vortices A and B can also be observed. The vortex A is the main recirculation vortex, which is accompanied by a small, counterrotating satellite vortex B, in the corner. In the case of combustion, when burned gases get produced, both vortices grow as shown in the second sketch of Figure 6.1. The growth of the recirculation vortices causes vortex A to be pushed downstream, while vortex B increases in size. As a consequence, smaller, counterclockwise rotating vortices C and D are created. This is illustrated in the third sketch. When vortex B becomes larger (as illustrated in the fourth sketch), vortex C may be forced upstream, causing a flow reversal on the edge of the step, tripping the boundary

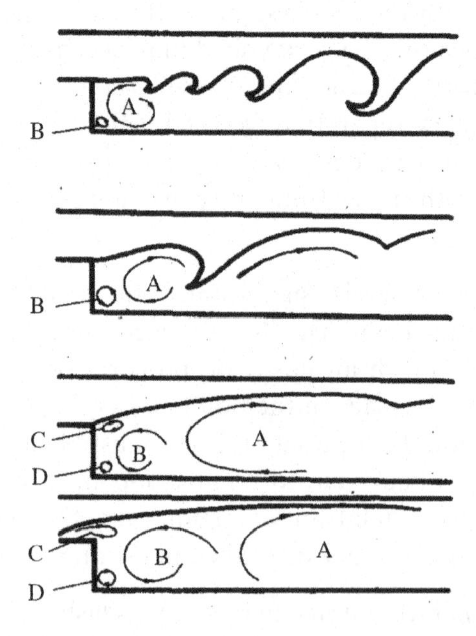

Fig. 6.1. Qualitative illustration of interaction between the recirculating and trailing vortices leading to flashback (Keller et al., 1982).

layer, and pushing the flame upstream. This was the main characteristic of the flashback observed (Keller et al., 1982).

Thibaut and Candel (1998) computationally analyzed the experiment of Keller et al. (1982) applying large eddy simulations (LES). They pointed out that the flashback observed by Keller et al. (1982) occurred for very strong oscillations. Assuming that linearized acoustics still provides a correct order of magnitude, a velocity fluctuation u' associated with a pressure fluctuation p' can be estimated from:

$$u' = \frac{p'}{\rho c} \tag{6.1}$$

For the measured (Keller et al., 1982) p' of 7 kPa for flashback, a fluctuational velocity of $u' \sim 17.5$ m/s was obtained (6.1) (Thibaut and Candel, 1998). This is much larger than the mean inlet velocity in these experiments, which was $u_{IN} = 13.3$ m/s. On a one-dimensional basis, this estimation would imply flashback by a periodic flow reversal at the inlet. This approximate comparison indicates that indeed very strong oscillations were leading to flashback. Related numerical simulations of Najm and Ghoniem (1994) showed how a downstream excitation over low-frequency pressure oscillations near the natural frequency of the combustion system reinforces the instabilities up to the point of flashback. High-frequency oscillations were not effective in that respect.

Dynamics and flashback characteristics of premixed hydrogen-enriched methane–air flames were investigated by Tuncer et al. (2006, 2009) experimentally, and analytically (based on a simple linearized acoustic model) on a test combustor. The basic design of the fuel–air premixing section of the latter represents a generic configuration with characteristic features similar to an industrial gas turbine, where the fuel is injected into the swirling cross-flow and mixes within a downstream distance before reaching the dump plane. The fuel (methane) was gradually enriched by hydrogen. Depending on the hydrogen content, two distinct regions of dominant mode, one with higher frequency (around 90 Hz) and another one with lower frequency (around 30–50 Hz), were identified. This transition occurred around 20–25% hydrogen content. The flashback was associated with periodic low-frequency fluctuations and instability occurring at rather low frequencies, with the associated

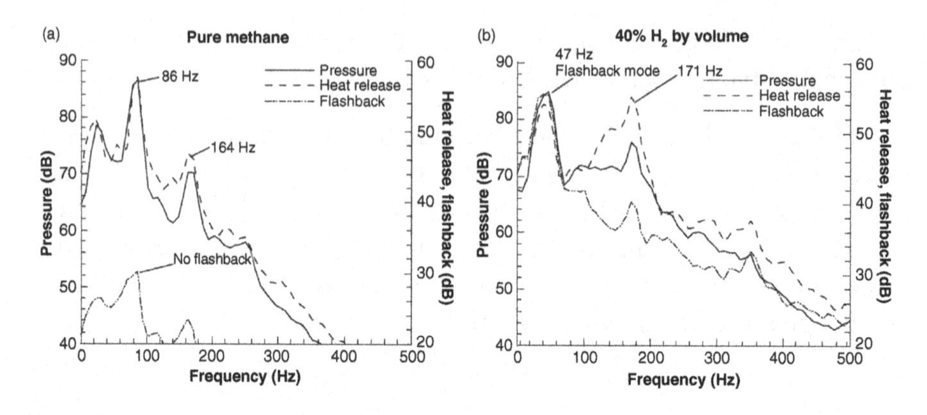

Fig. 6.2. Pressure, heat release, and flashback spectra ($\phi = 0.7$): (a) pure methane and (b) 40% hydrogen (Vol.) (Tuncer et al., 2009).

longer periods of oscillations, giving the flame more time to propagate upstream.

Figure 6.2 illustrates the spectra of pressure, heat release, and flashback signals. For pure methane, there is no flashback. Although the flashback detector registers a signal, its amplitude is low, and is at the noise level associated with the diagnostics. At 40% hydrogen content, the flashback signal at 47 Hz is clearly evident. The flashback activity and change in the dominant frequency in the pressure oscillations accompany one another, and both are at the same frequency. This means that as the flame moves back and forth inside the premixer and generates heat release fluctuations at the flashback frequency, the resulting heat release dynamics couples with an acoustic mode of the combustor system exciting this mode.

It was, thus, concluded (Tuncer et al. 2006, 2009) that in order to prevent the occurrence of flashback, attention should be focused primarily on eliminating the low-frequency pressure fluctuations. As the flashback was observed to be primarily triggered by thermoacoustic instability, an active control scheme designed to suppress the amplitude of thermoacoustic limit cycle oscillations was seen as a measure to resolve the flashback issue.

CHAPTER 7

Flashback Due to Turbulent Flame Propagation in the Core Flow

Flame propagation can take place in the wall boundary layers or in the free stream i.e. the core flow. Wall boundary layer flashback will be discussed separately in the following chapter.

As already discussed in Chapter 2, turbulent flame speed is not only a property of the mixture, but also of the turbulence. This results in a significant complication as for example, even the definition of the turbulent flame speed is afflicted with uncertainties. Lipatnikov and Chomiak (2002) discuss how the measurement of turbulent flame speed depends on different parameters, including the choice of the reference surface. Driscoll (2008) defined three types of turbulent flame speed: the global consumption speed, the local consumption speed, and the local displacement speed, which are not necessarily equal and show differences depending on the configuration. In many existing measurements (or models) the type of turbulent flame speed referred to was frequently not distinguished, which may be one of the causes of scatter observed in the data reported in the literature. In the present work, we will also refer to the existing correlations on the turbulent flame speed, as they are, without attempting to differentiate between different definitions.

Initial considerations on the calculation of the turbulent flame speed go back to Damköhler (Libby and Williams, 1994), the theories of whom are extended and improved by Shchelkin, Zimont, and, later on, many other researchers (Lipatnikov and Chomiak, 2002). For the flamelet regime, Damköhler postulated that the increase of the flame speed is due to the increase of the flame surface area by the wrinkling action. This resulted in the relationship $S_T/S_L = A_L/A_T \sim u'/S_L$, where A_L and A_T denote the instantaneous (wrinkled) and the time-averaged (smoothed) flame surface areas, respectively. For the thickened reaction zone, Damköhler hypothesized that the small-scale turbulence increases the transport rates and in analogy to the scaling relation for the laminar

Flashback Mechanisms in Lean Premixed Gas Turbine Combustion

burning velocity, the turbulent flame speed can be related to the laminar flame speed by $S_T/S_L = ((u'/S_L)(l_0/\delta_L))^{1/2}$. According to these modeling conceptions, which provided a basis for many further works, the following correlation can be expected to hold (Driscoll, 2008):

$$\frac{S_T}{S_L} = f\left(\frac{u'}{S_L}, \frac{l_0}{\delta_L}, \mathrm{Ma}_T\right) \tag{7.1}$$

In the above equation Ma_T is the turbulent Markstein number, which is neglected in many commonly used models. Recently, it is discussed (Driscoll, 2008) that the earlier functional relationship is still not sufficient, since the wrinkling process is "geometry dependent" and has a "memory" of whatever wrinkling occurred at upstream locations. Thus, it is argued (Driscoll, 2008) that additional parameters should be considered such as the nondimensional burner width or time, in Eq. (7.1). A detailed overview of models for the turbulent flame speed, at different levels of sophistication, can be found in the literature (Lipatnikov and Chomiak, 2002; Driscoll, 2008).

Many of the existing correlations are expressed in the following form:

$$\frac{S_T}{S_L} = 1 + A \cdot \mathrm{Re}_t^m \cdot \left(\frac{u'}{S_L}\right)^n \tag{7.2}$$

In such correlations, S_L is usually understood as the speed of an undisturbed (planar, unstretched) laminar flame, which is a characterizing property of the fuel–air mixture. The correlation suggested by Gülder (1990) is recovered by substituting, $A = 0.62$, $m = 0.25$, and $n = 0.5$ in (7.2), for the flamelet regime. The correlation suggested by Liu (1991) corresponds to, $A = 0.435$, $m = 0.44$, and $n = 0.4$.

In Figure 7.1, the measured (Abdel-Gayed et al., 1984) S_T curves (denoted as u_t in the figure) as a function of u' at different equivalence ratios for atmospheric propane flames are shown. One can see the bending of the S_T curves with increasing u', and their decline after having reached a maximum value, toward quenching (shadings in Figure 7.1 indicate quenching), since flame stretch effects and local quenching start to play a dominating role in this region.

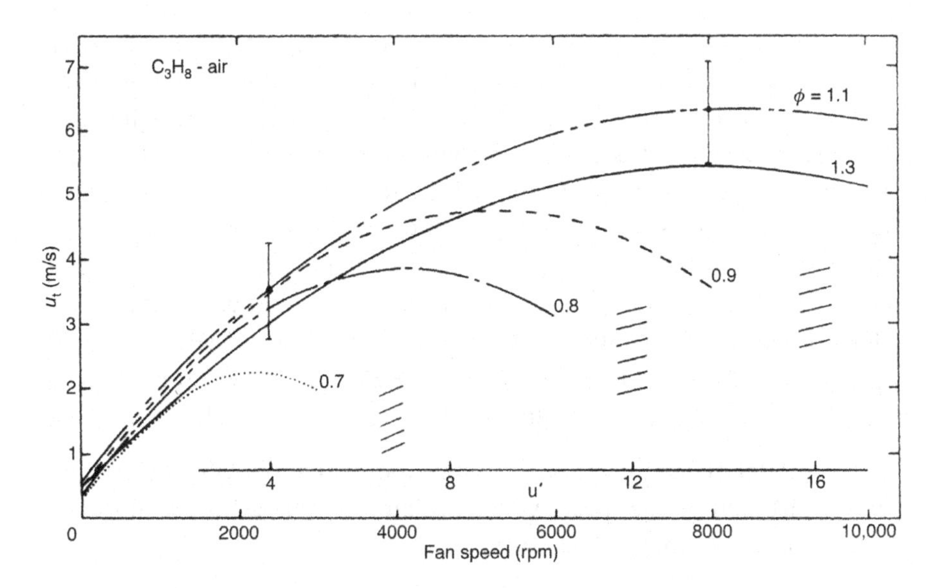

Fig. 7.1. Measured turbulent flame speed S_T *(denoted as* u_t *in the figure) as a function of* u' *at different equivalence ratios for atmospheric propane flames. Shading indicates the quenching regions (Abdel-Gayed et al., 1984).*

In a further experimental investigation, Abdel-Gayed and Bradley (1989) investigated the quenching behavior of turbulent premixed flames, where they suggested the following criterion for the complete quenching:

$$\mathrm{Ka} \cdot \mathrm{Le} = 1.5 \tag{7.3}$$

for flames with $\mathrm{Re}_\mathrm{T} > 300$. One can see in (7.3) that flames with low Le can sustain a higher level of stretch.

Correlations like (7.2) predict the bending behavior of the S_T curve with increasing u'. However, they do not embody any mechanism for taking care of quenching and decline of the S_T curve at high u'. The predicted $S_\mathrm{T}/S_\mathrm{L}$ monotonically increases with increasing u'/S_L, however, with always decreasing but positive slope.

The correlation suggested by Zimont et al. (2001), which covers the flamelet and thickened wrinkled flame regions, calculates the turbulent flame speed using the following expression:

$$S_\mathrm{T} = AG\alpha^{-1/4} l_0^{1/4} u'^{3/4} S_\mathrm{L}^{1/2} = AGu' \,\mathrm{Da}_\mathrm{T}^{1/4} \tag{7.4}$$

In the above expression, A is a model constant with a suggested value of A = 0.52. The thermal diffusivity α is based on unburned mixture conditions. G is the so-called "stretching factor," through which the model can predict decline of S_T with increasing u', due to dominating stretch effects. In Zimont's model, the strain rate is modeled through the dissipation rate $\varepsilon\,[\varepsilon = C_D(u'^3)/l_0$, with $C_D = 0.37]$ of turbulence kinetic energy k. It is assumed that the stretch effects and local quenching start to play a role if the dissipation rate ε exceeds a certain critical value ε_{cr}, which represents the dissipation rate at the critical rate of strain g_{cr}, with $\varepsilon_{cr} = 15\nu g_{cr}$. The latter is modeled by $g_{cr} = BS_L^2 / \alpha$, where B is a model constant with the suggested value of 0.5.

The model of Schmid et al. (1998) proposes the following expression for the turbulent flame speed, which is claimed to be valid for all turbulent premixed combustion regimes:

$$\frac{S_T}{S_L} = 1 + \frac{u'}{S_L}(1 + \mathrm{Da}_T^{-2})^{-(1/4)} \tag{7.5}$$

According to this model, the effect of Da_T declines quite rapidly with increasing Da_T. At large Da_T ($\mathrm{Da}_T \gg 1$), the limiting case of Damköhler for large-scale turbulence is recovered: $S_T/S_L = 1 + (u'/S_L)$. For the other limiting case of $\mathrm{Da}_T \ll 1$, using $\mathrm{Re}_T = \mathrm{Da}_T(u'/S_L)^2$, in the limit, the relationship $S_T / S_L = 1 + \mathrm{Re}_T^{1/2}$ is obtained, which resembles Damköhler's correlation for small-scale turbulence.

Predictions for the likeliness of flashback due to turbulent flame propagation in the core can be made using the previous expressions. In practical systems, Da_T values are usually large. Thus, the limiting behavior of Eq. (7.5) for $\mathrm{Da}_T \gg 1$, i.e., the expression $S_T/S_L = 1 + (u'/S_L)$, can be used for a quick estimation (Lieuwen et al., 2008b). Although flashback can locally occur along any streamline, where the turbulent flame speed exceeds the flow velocity, we consider for this estimation a radially homogeneous flow field. Expressing u' by means of a turbulence intensity I based on the bulk axial velocity U, i.e., $u' = IU$, leads to $S_T = S_L + IU$ in the large Da_T limit of Eq. (7.5). Thus, for preventing flashback, i.e., for $U > S_T$, the condition $U > S_L/(1 - I)$ must hold. Flows in premixers are normally highly swirled and the swirl component contributes considerably to the turbulence kinetic energy. Still, turbulence intensities based on the bulk axial velocity do not usually exceed 20% (I = 0.2), leading to the following approximate flashback limit for the axial bulk velocity (U_{FL}):

$$U_{\mathrm{FL}} \approx 1.25 S_{\mathrm{L}} \tag{7.6}$$

which should not fall below this value for preventing flashback due to turbulent flame propagation in the core flow. Assuming a rather high unburned mixture temperature of 800 K, the estimated approximate flashback limits (U_{FL}), based on Eq. (7.6), for equivalence ratios 1 and 0.5 are provided in Figure 7.2 for methane–air and hydrogen–air mixtures as a function of pressure.

The indicated values (Figure 7.2) for hydrogen are approximately six to seven times higher than those for methane, which emphasizes the critical role of hydrogen. With increasing equivalence ratio from ϕ = 0.5 to 1.0, the values increase by a factor of approximately 2. The curves decline with increasing pressure (Chapter 2). The bulk axial flow velocities encountered in gas turbine combustor premixers are much

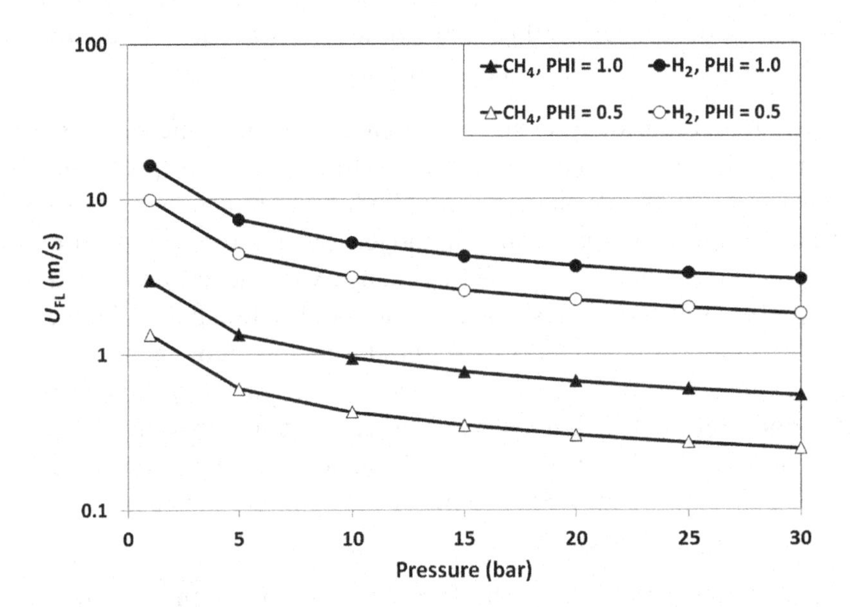

Fig. 7.2. Estimated approximate flashback limits due to turbulent flame propagation in the core, in terms of the bulk velocity (U_{FL}), after Eq. (7.6), for methane–air and hydrogen–air mixtures, assuming an unburned mixture temperature of T_u = 800 K, for equivalence ratios ϕ = 1 and 0.5, as a function of pressure.

higher, usually above approximately 50 m/s (Lieuwen et al., 2008b), than the limits indicated in Figure 7.2. Thus, the previous estimation (Figure 7.2) implies, as a first estimate, that flashback due to turbulent flame propagation in the core is less likely to occur in practical applications. This is especially true for natural gas. For hydrogen-containing fuels, the safety margin is smaller. In reality, there are, of course, departures form the assumed ideal situation. Inhomogeneities in the velocity field with low-velocity regions and/or extremely high turbulence intensities (e.g., in the wake regions of swirler vanes, fuel jets) increase the flashback propensity. In lean mixtures, the flame speeds are lower as seen in Figure 7.2 in the comparison of $\phi = 1$ and 0.5 curves. In fuel-rich mixtures ($1 < \phi < 2$) the laminar flame speed attains its maximum, which can be substantially higher than the stoichiometric value, especially for hydrogen (Chapter 3). Thus, deficient mixing is a further factor increasing the flashback propensity. Sufficiently high axial velocities along with preferably homogeneous distributions of axial velocity and composition fields are, thus, the major design criteria for preventing flashback. Too high turbulence intensities should also be avoided. This can, e.g., be achieved by a slight reduction of the swirl number, as the swirling motion considerably contributes to turbulence production. However, since this can cause a deterioration in the other features such as the mixing quality, it is critical to find an optimum.

The foregoing analysis suggests that turbulent flame propagation flashback is unlikely to occur in gas turbine applications (provided that design conditions exhibit sufficiently high and uniform velocity distributions along with a preferably homogeneous mixture composition in the premixer). This is supported by many experimental investigations. Wohl (1953) observed flashback at comparably low flow velocities in nonswirling pipe flows, namely, at about 3 m/s for propane flames and in the range 5–10 m/s for hydrogen-containing fuels. In the review of Plee and Mellor (1978) on flashback in gas turbines utilizing hydrocarbon fuels, no evidence of classical flashback, i.e., by turbulent flame propagation, was reported, except for catalytic combustors with very low reference velocities.

An interesting aspect is the fact that the condition of maximum turbulent flame speed is generally not much far away from the extinction limit. This can be seen, e.g., in Figure 7.1. Thus, although too high turbulence intensities are generally not desirable in order to limit the turbulent flame speed, and, thus, the flashback propensity, an increase of the turbulence intensity (which corresponds to a decrease in Da_T) beyond a certain level can reduce flashback propensity, since an upstream flame propagation will then not be possible due to quenching. This technique can also be utilized for preventing flashback.

Daniele and Jansohn (2012) have recently proposed a correlation for turbulent flame speed for $CH_4/CO/H_2$ mixtures, which was covering a wide range of operation conditions (pressures up to 2.0 MPa, inlet temperatures up to 773 K) and fuel compositions (mole fractions of CO and H_2 up to 0.67). The correlation obtained on an optically accessible high-pressure combustion chamber with a thermal power of 0.5 MW reads as follows:

$$\frac{S_T^{0.05}}{S_L} = 337.45\left(\frac{u'}{S_L}\right)^{0.63}\left(\frac{l_0}{\delta_L}\right)^{-0.37}\left(\frac{p}{p_R}\right)^{0.63}\left(\frac{T}{T_R}\right)^{-0.63} \tag{7.7}$$

where the reference pressure and temperature are set to be $p_R = 0.1$ MPa and $T_R = 1$ K.

As far as the hydrogen blend fuels are concerned, an additional aspect shall be addressed. Hydrogen addition to a hydrocarbon (e.g., methane) fuel causes an increase of the laminar flame speed S_L. An accompanying increase in the turbulent flame speed S_T is modeled in the previous correlations (Eqs. (7.2), (7.4), (7.5), and (7.7)), as S_T depends on S_L. Under the assumption of an unaltered turbulence field (u'), an expression such as Eq. (7.2) predicts an underproportional increase of S_T compared with the increase of S_L, since (u'/S_L) decreases with increasing S_L. However, it was experimentally observed that the increase in S_T may be stronger than the increase in S_L if the latter is caused by hydrogen addition. Such effects are normally attributed to the role of Lewis number.

Experimental results of Kido et al. (2002) are shown in Figure 7.3, where the measured turbulent flame speeds in a constant-volume vessel are displayed as a function of turbulence intensity for different fuel compositions and equivalence ratios. The H07-15N and M07-15N curves represent nitrogen-diluted hydrogen–air and methane–air flames, respectively. Both of them have the same equivalence ratio (0.7) and practically

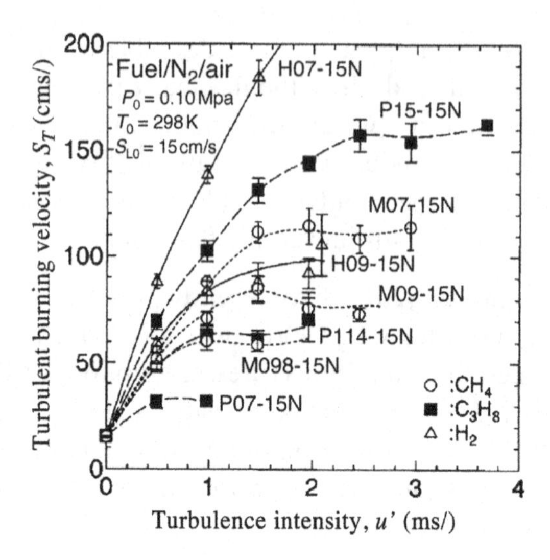

Fig. 7.3. Variation of turbulent burning velocity with turbulence intensity for different fuel compositions and equivalence ratios (Kido et al., 2002).

the same laminar flame speed (15.25 m/s and 15.10 cm/s, controlled by the nitrogen content), and result in quite similar Re_T and Da_T values at the same turbulence intensity. However, they exhibit different Lewis numbers, which has the value of 0.41 for the hydrogen flame (H07-15N) and 0.89 for the methane flame (M07-15N). One can see that the turbulence flame speed measured for H07-15N lies much higher compared than that for M07-15. At low turbulence intensities, the difference is quite small but increases with increasing u', and amounts approximately 40% for $u' = 1.5$ m/s. This finding supports the thesis of the strong influence of the Lewis number on turbulent flame speed.

A more recent experimental investigation was presented by Guo et al. (2010) on hydrogen-enriched turbulent lean premixed methane–air flames using a V-shaped flame configuration. The variations of the measured ratio of the turbulent flame speed to the laminar flame speed with increasing hydrogen content of the fuel for two equivalence ratios (0.55, 0.60) and a constant turbulence intensity of 4% are displayed in Figure 7.4 (Guo et al., 2010).

It is observed (Figure 7.4) that when the fraction of hydrogen increases at a constant equivalence ratio and a similar turbulence intensity, the ratio of turbulent to laminar flame speeds also increases, although u'/S_L

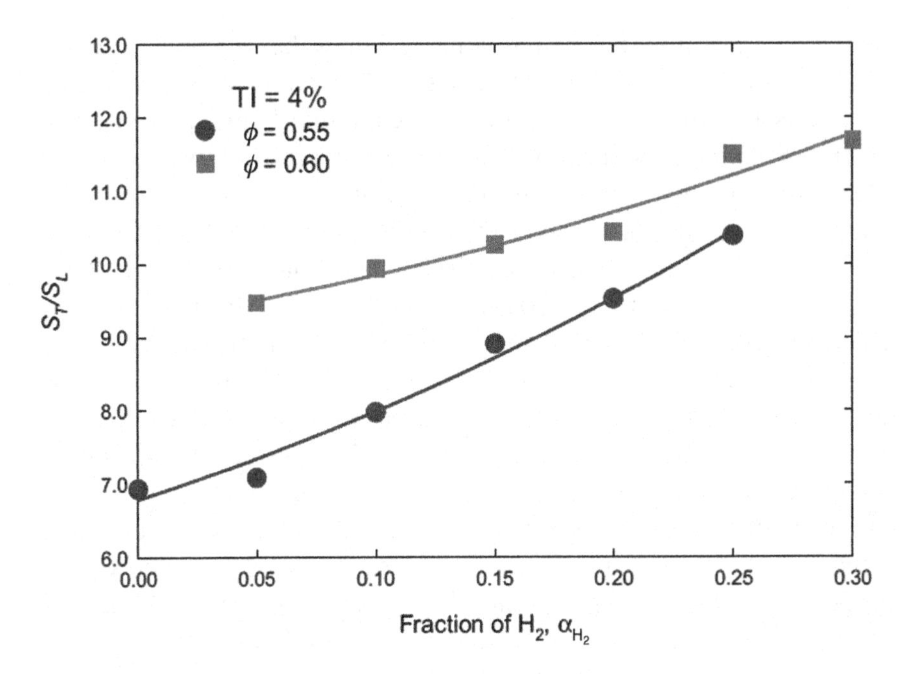

Fig. 7.4. Variation of S_T/S_L *with hydrogen fuel molar fraction in a hydrogen-enriched premixed methane–air flame for a turbulence intensity of 4%, for equivalence ratios 0.55 and 0.60 (Guo et al., 2010).*

decreases due to the increase in S_L. This indicates that with hydrogen enrichment, turbulent flame speed increases faster than the laminar flame speed. This also indicates the importance of the Lewis number effect in interaction with the turbulent flow field in addition to purely kinetic effects. The further analysis of the flame surface characteristics confirmed an increase of the total flame surface area by hydrogen addition, where a slight decrease in the local flame surface density was overcompensated by the increase in the flame brush thickness. It was also argued that the stretched local laminar burning velocity could be enhanced by hydrogen addition. As both effects are due to a decrease in the Lewis number by hydrogen addition, the results confirm the importance of Lewis number effects on the flame speed in turbulent premixed flames.

As mentioned before, some researchers attribute this behavior to the Lewis number effects, which were outlined shortly in Chapter 2. Since hydrogen possesses a large mass diffusivity, leading to the comparably low Lewis numbers, it can diffuse much more rapidly to the flame zone, leading to a local hydrogen enrichment and shifting of the local mixture composition. For planar laminar flames the combined reaction and heat and mass transport processes, and therefore, any Lewis number effects are already incorporated in the laminar flame speed S_L. The wrinkling of the flame front by turbulence activates, however, a further effect associated with the Lewis number. For a curved flame front, the condition Le < 1 leads to enhanced reactivity (increased local and temporal laminar flame speed) in positively curved flame fragments (convex toward the unburned mixture), while the reactivity is reduced in negatively curved parts. One may argue that for a strongly wrinkled flame in a highly turbulent flow, both effects should cancel out. Here, the so-called "leading edge concept" (Lipatnikov and Chomiak, 2002) postulating that the leading (convex) parts of the turbulent flame brush determine the overall flame speed, may be referred to explain the dominating effect of the convex parts, i.e., the increase of the overall flame speed. Furthermore, increasing and decreasing local flame speeds in the convex and concave parts lead to an increase of the wrinkling, the so-called thermodiffusive instability with self-induced wrinkling and growth of the flame surface, which can be seen as a source of the overproportional increase of the turbulent flame speed. Flame stretch effects can also contribute to this behavior, as a decrease in fuel Lewis number may cause an enhancement of the local and instantaneous laminar flame speed of the stretched flame sheet. According to Dinkelacker et al. (2011), the thermodiffusive instability and self-induced wrinkling effects are important not only at low turbulence levels but also in a highly turbulent flow.

In order to account for the Lewis number effects, further correlations for the turbulent flame speed are proposed, with explicit reference to the Lewis number. Bradley (1992) suggested the following correlation:

$$\frac{S_T}{S_L} = 1 + \frac{0.15}{\mathrm{Ka} \cdot \mathrm{Le}} \left(\frac{u'}{S_L} \right)^2 \tag{7.8}$$

where he used the following definition of the Karlovitz number:

$$\mathrm{Ka} = \frac{0.157}{\mathrm{Re}_T^{0.5}} \left(\frac{u'}{S_L} \right)^2 \tag{7.9}$$

Guo et al. (2010) suggested the following correlation, which showed a very good agreement to their data:

$$\frac{S_T}{u'} = \frac{1.00554}{(\mathrm{Ka} \cdot \mathrm{Le})^{0.5953}} \tag{7.10}$$

using the Karlovitz number definition of Bradley (Eq. (7.9)).

Within the framework of their computational analysis using a reaction progress variable–based algebraic flame surface wrinkling combustion model, Dinkelacker et al. (2011) suggested the following relationship, where pressure effects were also incorporated (p_0= 1 bar):

$$\frac{S_T}{S_L} = 1 + \frac{0.46}{\mathrm{Le}} \mathrm{Re}_T^{0.25} \left(\frac{u'}{S_L}\right)^{0.3} \left(\frac{p}{p_0}\right)^{0.2} \tag{7.11}$$

They additionally emphasized the importance of the procedure for determining the effective mixture Lewis number for fuel blends. For a methane–hydrogen fuel blend, the effective mixture Lewis number (Le) may be approximated based on the Lewis numbers of methane (Le_{CH_4}) and hydrogen (Le_{H_2}). A model proposed by Law and Kwon (cited in Dinkelacker et al., 2011) suggests a heat release weighted average of the methane and hydrogen Lewis numbers (Model A in Dinkelacker et al., 2011). An alternative approach is a mole fraction weighted average (Model B). In contrast to these approaches, Dinkelacker et al. (2011) suggest a Le Chatelier rule–like weighting (Model C) given as follows:

$$\frac{1}{\mathrm{Le}} = \frac{X_{CH_4}}{\mathrm{Le}_{CH_4}} + \frac{X_{H_2}}{\mathrm{Le}_{H_2}} \tag{7.12}$$

Experimental and predicted flame shapes for pure methane and for a hydrogen–methane fuel blend (40% H_2 + 60% CH_4) for p = 0.5 MPa and ϕ = 0.5 are compared in Figure 7.5 (Dinkelacker et al., 2011).

In the predictions two models for determining the effective mixture Lewis number were used (Models A and C). One can see that both models agree with each other and with the experiments quite well for the case of pure methane flame. However, for the hydrogen–methane fuel blend (40% H_2 + 60% CH_4) the procedure for determining the effective

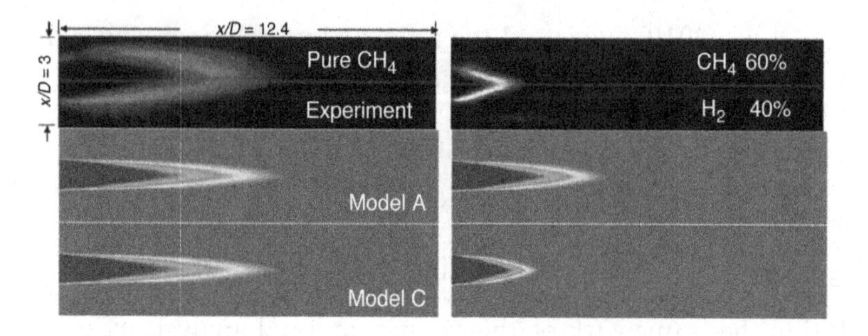

Fig. 7.5. Experimental and predicted flame shapes for pure CH_4 and 40% H_2 + 60% CH_4 for p = *0.5 MPa and* $\phi = 0.5$ *(Dinkelacker et al., 2011).*

mixture Lewis number plays a role. The heat release weighting (Model A) does not show a good agreement with the measurements, whereas the suggested Le Chatelier rule–like weighting (Eq. (7.12), Model C) shows a much better agreement with experiments.

A much deeper discussion on the speed and structure of turbulent flames including the still not completely understood thermodiffusion and Lewis number effects can be found in Lipatnikov and Chomiak (2002) and Driscoll (2008).

On the other hand, there are also measurements that do not predict the above-mentioned behavior. Liu and Lenze (1988) investigated turbulent flame speeds of methane–hydrogen flames at various turbulence levels. They found that S_T remained rather uninfluenced by the fuel composition, provided S_L was maintained constant. Thus, the turbulent velocities could be expressed by the same correlation (Eq. (7.2)) for methane and hydrogen turbulent premixed flames. Mandilas et al. (2007) experimentally investigated the influence of hydrogen addition on laminar and turbulent flame speeds using spherical expanding flames. They also observed that the ratio of turbulent to laminar burning velocities did not significantly change with the addition of hydrogen at similar turbulence conditions. Schmid et al. (1998) could predict turbulent flame speeds of premixed methane and hydrogen flames using the same correlation (Eq. (7.5)) equally well, implying, again, a subordinate role of fuel composition other than its impact on the laminar flame speed. On the other hand, the influence of fuel composition in the case of hydrogen was confirmed in many further experimental and computational investigations. Hawkes and Chen (2004) showed by means of a direct numerical simulation analysis that turbulent flame speed increased faster than the laminar flame speed by hydrogen addition to the fuel for a lean premixed methane–air mixture at the same turbulent condition. The experimental investigation of Cohé et al. (2007) demonstrated that the addition of hydrogen to the fuel for a methane–air turbulent premixed flame resulted in an increase in fractal dimension, which also implied that the addition of hydrogen at a similar turbulence intensity led to an increase in the ratio of turbulent to laminar burning velocities, based on the fractal theory. These dissonant findings in literature imply that the current understanding on the effect of hydrogen blending of hydrocarbon fuels in turbulent premixed combustion is not complete. Thus, there is a need for further investigation in this field.

CHAPTER 8

Flashback Due to Flame Propagation in Boundary Layers

In the boundary layers over solid walls, the flow velocity is gradually reduced toward the wall, due to the no-slip wall boundary condition. Thus, even for high freestream velocities, it is possible that the flame speed outbalances local flow speed in the boundary layer, at a particular distance from the wall. However, flashback potential is counteracted by flame quenching due to heat loss to the wall, and flame stretch. As also discussed by Eichler (2011), several possible scenarios for wall boundary layer flashback in gas turbine burners are illustrated in Figure 8.1.

The scenario illustrated in Figure 8.1a, representing flashback in a mixing tube with unswirled flow, corresponds to the "classical" unidirectional boundary layer flow (e.g., boundary layers on flat plates, pipe walls) and can be analyzed correspondingly. In the remaining scenarios, the interaction of the boundary layer with the wakes behind the fuel injection jets (Figure 8.1b) and with the swirling flow (Figure 8.1c and d) leads to a more complex boundary layer structure.

8.1 FLAME PROPAGATION IN LAMINAR BOUNDARY LAYERS

Flashback in laminar boundary layers has been a classical topic that has been extensively investigated (Lewis and von Elbe, 1987). The proposed "critical velocity gradient" concept, illustrated in Figure 8.2, has become a widely adopted model for the quantification of wall flashback.

For a stable flame front at a distance δ_b from the wall (Figure 8.2), the local flow velocity and the flame speed (S_f) need to be in balance. Assuming a linear velocity profile in the vicinity of the wall (denoting the wall shear stress by τ_W) one can write the following:

$$g_C = \left.\frac{\partial u}{\partial y}\right|_{y=0} = \frac{\tau_W}{\mu} = \frac{S_f|_{y=\delta_b}}{\delta_b} \tag{8.1}$$

Flashback Mechanisms in Lean Premixed Gas Turbine Combustion

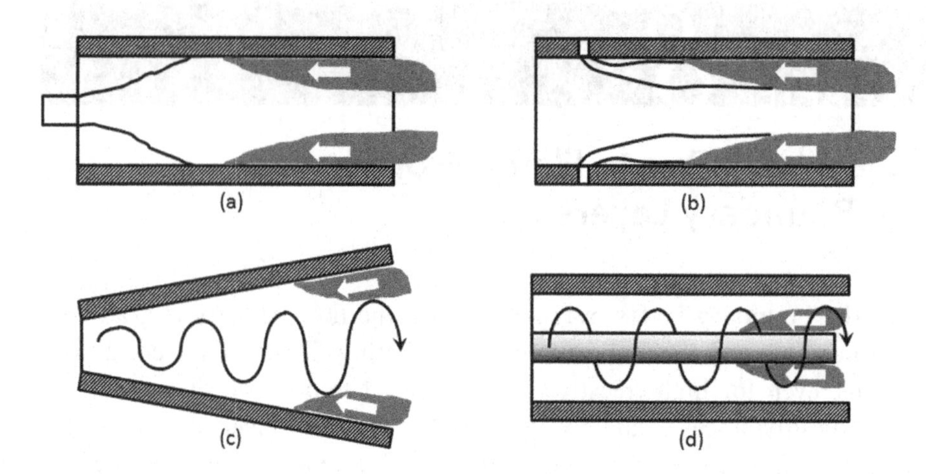

Fig. 8.1. Illustration of several wall boundary layer flashback scenarios in gas turbine burners.

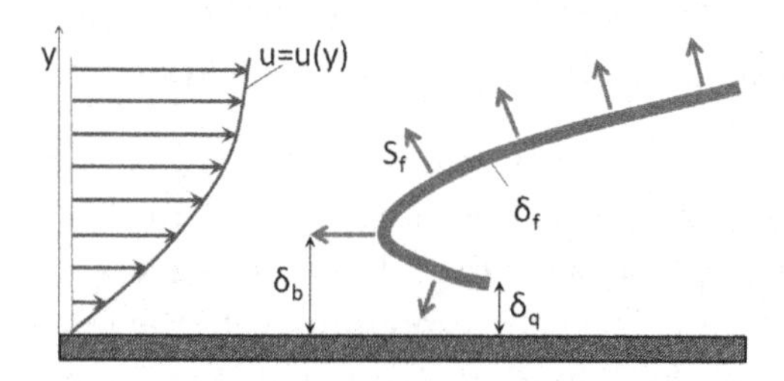

Fig. 8.2. Critical velocity gradient concept (δ_q, quenching distance; δ_b, penetration distance; δ_f, flame thickness; S_f, flame speed; T_w, wall temperature; u(y), boundary layer velocity profile).

where g_C is the so-called "critical velocity gradient," which generally depends on fuel-oxidizer kinetics, ϕ, T, p, and T_W.

The velocity gradient at the wall (i.e., the wall shear stress) can be related to the freestream velocity, if the shape of the velocity profile can be estimated with sufficient accuracy. For fully developed laminar pipe flow (Poiseuille flow), using the analytical parabolic velocity profile, the velocity gradient can be obtained as a function of bulk flow velocity (U) and pipe diameter (d) as follows:

$$g = 8\frac{U}{d} \tag{8.2}$$

Eq. (8.1) is not of much practical use, since the penetration depth (δ_b) is normally not known, and cannot be easily predicted. The same also applies to the near-wall flame speed (S_f), due to curvature, and stretch effects as well as thermal and radical quenching. The penetration depth generally correlates with the quenching distance, which does, however, not necessarily scale with the tube diameter. These uncertainties necessitate empirical approaches for the determination of g_C, where the universality becomes a problem. Assuming that the penetration depth is of the order of the flame thickness and S_f can be approximately represented by the planar, unstretched laminar flame speed S_L, the following proportionality can be obtained:

$$g_C \propto \frac{S_L^2}{\alpha} \tag{8.3}$$

which underlines the substantial role of the laminar flame speed. One can deduce from Eq. (8.3) that the propensity for boundary layer flashback would rapidly increase with increasing hydrogen content of the fuel.

Dam et al. (2011b) experimentally investigated the flashback propensity of syngas fuels. As different flashback mechanisms were discussed, the main focus was on the boundary layer flashback in straight tubes under laminar flow conditions.

Figure 8.3 shows the measured (Dam et al., 2011b) critical velocity gradients of different fuel mixtures at different values of the volumetric percent fuel (%F) in the air–fuel mixture.

One can see that the boundary layer flashback propensity (indicated by higher values of g_F) increases with increasing hydrogen fractions in the fuel mixture. For H_2–CO mixtures (Figure 8.3a), the effect of hydrogen addition is more significant at rich conditions, whereas the H_2–CH_4 mixtures (Figure 8.3b) exhibit a higher sensitivity for lean conditions.

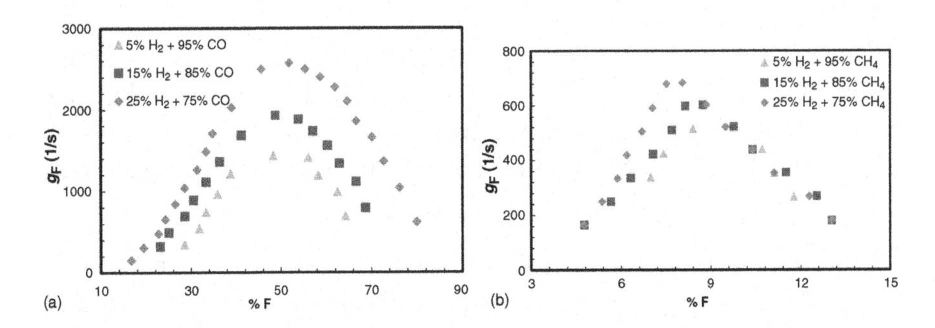

Fig. 8.3. Measured critical velocity gradients (g_F) for different fuel compositions as a function of volume percent fuel (%F): (a) H_2–CO blends and (b) H_2–CH_4 blends (Dam et al., 2011b).

Dam et al. (2011b) measured the critical velocity gradient of various mixtures of H_2–CO and H_2–CH_4 (up to 25% H_2) under three external forcing frequencies (100, 300, and 700 Hz) to develop a qualitative understanding of the combustion instability effect on critical velocity gradients. No remarkable effect of the external excitation on the flashback propensity was observed for the investigated experimental setup (Dam et al., 2011b).

8.2 FLAME PROPAGATION IN TURBULENT BOUNDARY LAYERS

For turbulent boundary layers, in many approaches, the concept of critical velocity gradient for laminar boundary layer flashback is transferred to turbulent flow (Bollinger, 1958), where the wall shear stress needs to be calculated from a suitable expression for turbulent flow instead of Eq. (8.2). For cases where flashback takes place within the laminar sublayer ($\delta_b^+ \leq 5$), the analogy, which relies on the comparison of the flow velocity with the laminar flame speed, is quite reasonable, provided that the wall shear stress is estimated adequately. In turbulent flow, the wall shear stress can be estimated by an empirical expression for fully developed turbulent pipe flow such as the Blasius correlation (Schlichting, 1979), where the velocity gradient (g) at the wall is given as follows:

$$g = \frac{\tau_w}{\nu} = 0.03995 U^{7/4} \nu^{-(3/4)} d^{-(1/4)} \tag{8.4}$$

One can see that g depends much strongly on the velocity than on the diameter, which is also stronger compared with the laminar flow (where $g \sim U$).

Eichler (2011) presented a comparison of critical velocity gradient values for atmospheric premixed hydrogen–air flames in tube burners with varying diameters, obtained by different authors under laminar (Elbe and Mentser, 1945) and turbulent (Khitrin et al., 1965; Fine, 1958) flow conditions, which is shown in Figure 8.4.

One can see that the turbulent-to-laminar critical gradient ratio is about 3, i.e., $g_{C,T}/g_{C,L} \sim 3$ at maximum flashback propensity around $\phi \sim 1$–1.5, whereas this ratio varies for different values of the equivalence ratio. Wohl (1953) reported the same ratio ($g_{C,T}/g_{C,L} \sim 3$) at $\phi \sim 1.1$

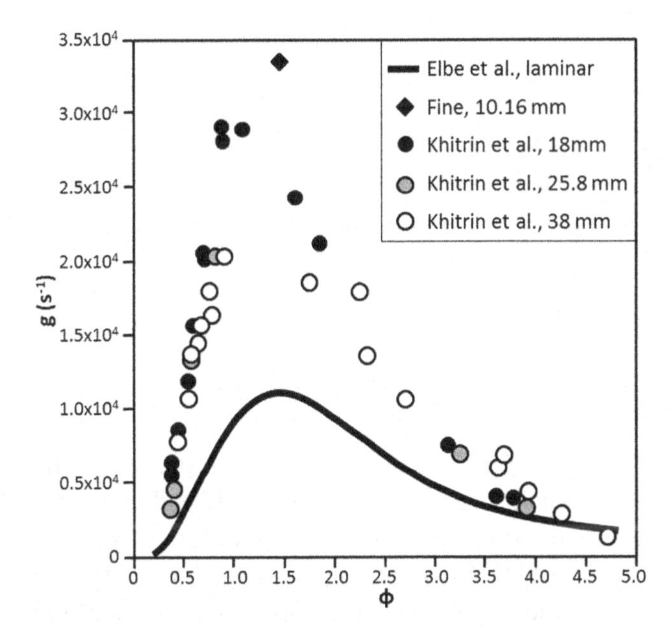

Fig. 8.4. Critical velocity gradients as a function of equivalence ratio for laminar and turbulent hydrogen–air flames in unconfined tube burners at atmospheric conditions.

for propane–air mixtures. A possible interpretation (Wohl, 1953) for $g_{C,T}$ being larger than $g_{C,L}$ is that the flashback occurs outside the laminar sublayer, where turbulence increases the flame speed. On the other hand, there are experimental investigations (Bollinger and Edse, 1956; Bollinger, 1958) that indicate that the penetration depth remains rather within the laminar sublayer. Therefore it remains unclear if the increase in g_C results from a decrease in δ_b, an increase in S_f, or both effects.

Earlier boundary layer flashback studies are based on tube burners, which held the flame in a free environment or in a chamber with a remarkable jump in cross-section. For gas turbine burners, this "unconfined" configuration represents rather a special case, which is, however, not representative of the situation, where the flame enters the premixing duct through the boundary layer (e.g., due to an intermittent velocity drop) and has to be washed out. Eichler and Sattelmayer (2010) and Eichler et al. (2011) elaborated on this point and performed tests in "confined" arrangements, where the flame was already stabilized by pilot burners downstream of the boundary layer prior to flashback

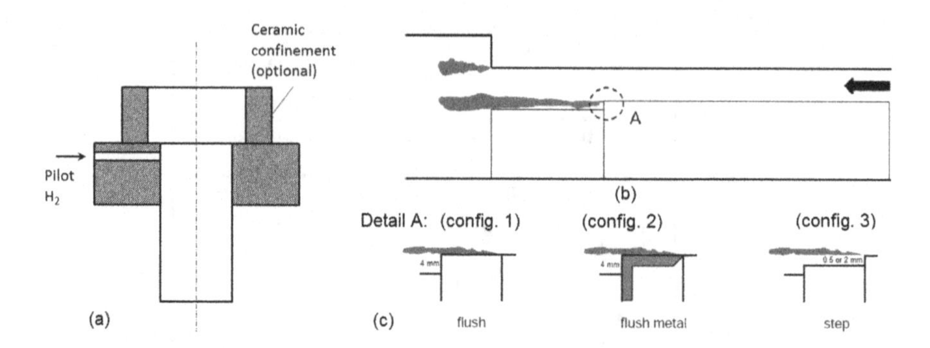

Fig. 8.5. Experimental configurations used by Eichler et al.: (a) tube burner, (b) channel section, and (c) detail A.

(Figure 8.5). For the tube burner (Figure 8.5a), the tube with and without the ceramic block correspond to "confined" and "unconfined" cases, respectively. For the channel (Figure 8.5b), different configurations are applied at the end of the 595 mm long lower wall. Velocity fields were obtained from PIV and CFD calculations using a two-equation turbulence model. It was observed that the critical velocity gradient values delivered by the Blasius correlation (Eq. (8.4)) also provided a satisfactory agreement with measurements within an approximate error band of ±10%.

The results of the flashback experiments (Eichler et al., 2011) are illustrated in Figure 8.6. (Please note that in the present illustration, instead of the scattered data, trendlines are drawn for some experiments.)

Monotonically increasing critical gradients with increasing fuel equivalence ratio are observed for all configurations (Figure 8.6). One can see that the critical gradients for the unconfined tube burner lie substantially below all other configurations, the difference approaching almost one order of magnitude toward stoichiometry. The results for the channel with different configurations indicate lower flashback propensity with increasing step height. It can be seen in Figure 8.6 that the flashback limits for the confined tube burner follow the channel values. This implies that the increase in flashback propensity observed in the channel is not caused by some peculiarities of the setup, but due to a fundamental difference between confined and unconfined flameholding prior to flashback.

Eichler et al. (2011) explain the reduced flashback propensity of unconfined configurations by radial outflow of the fresh mixture that is

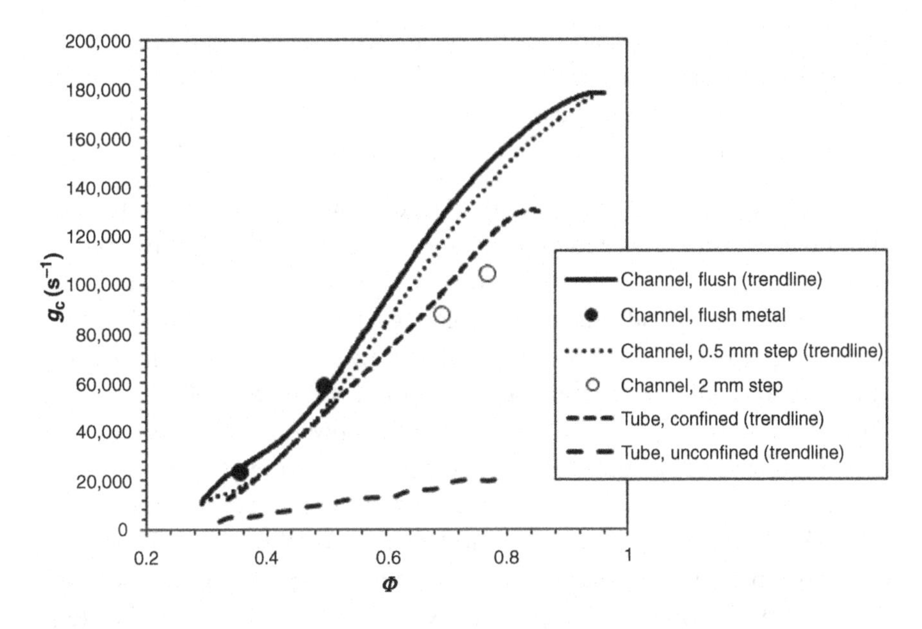

Fig. 8.6. Flashback limits for tube and channel configurations of turbulent hydrogen–air flames.

accelerated toward the quenching gap, caused by the rim, as well as the reduction of the flame-generated adverse pressure gradient by the pressure drop across the quenching gap. This situation is argued to reduce the flashback propensity for two reasons. On the one hand, the local adverse pressure gradient imposed on the boundary layer by the flame anchor is diminished by the pressure drop across the quenching gap. On the other hand, the flame is deterred from entering the boundary layer region due to the outward radial fluid motion. For a confined flame stabilized at a small backward-facing step (configuration 3), the physical picture is slightly different. Quenching causes a gap downstream of the step edge with an associated pressure drop in a similar way as for the unconfined case. However, the flow through this gap is obstructed by the offset channel wall. Thus, with decreasing step height, the fluid motion perpendicular to the main flow direction is retarded and the flame anchor moves closer to the boundary layer region. The quenching distance also decreases, since a decrease of the gap leakage flow reduces the convective heat loss from the flame preheat zone. These two effects result in an increased flashback propensity with decreasing step height as experimentally observed (Figure 8.6). A confined flame, stabilized flush with the duct wall (configuration 1) can hence be viewed as the upper limit for flashback propensity for a given turbulent boundary layer state. Thus, for a conservative design path, Eichler et al. (2011) suggest the consideration of their results obtained for confined flames with flush flameholding instead of open tube burner values.

Shaffer et al. (2012) performed a large number of flashback tests (at standard pressure and temperature) on a "confined" jet flame burner (consisting of a quartz burner tube confined by a larger quartz tube). Various fuel compositions of hydrogen, carbon monoxide, and natural gas were premixed with air at equivalence ratios corresponding to constant adiabatic flame temperatures (AFT) of 1700 and 1900 K. Once a flame was stabilized on the burner, the air flow rate was gradually reduced while holding the AFT constant via the equivalence ratio until flashback occurred. The covered ranges for the fuel composition were: AFT (K), 1700–1900; H_2 (vol%), 50–100; CO (vol%), 0–50; CH_4 (vol%), 0–50. As in previous studies (Bollinger and Edse, 1956), it was found that flashback conditions were greatly affected by the burner tip temperature. Thus, great care was paid to exclude such effects, by allowing enough time (20–30 min) after each test, for achieving thermal equilibrium at the burner tip. The velocity gradient was correlated to flow data through the Blasius correlation (8.4).

After a statistical analysis of the data, Shaffer et al. (2012) developed the following correlation (Shaffer, 2012) between the critical velocity gradient g_C (in 1/s), the composition of the $H_2/CO/CH_4$ fuel (in vol%), and the AFT (in K):

$$g_C = (2.401 \cdot H_2 + 5.318 \cdot CO - 1.632 \cdot CH_4 - 0.264 \cdot AFT + 0.0604 \cdot H_2 \cdot CH_4 + 0.00264 \cdot CH_4 \cdot AFT)^2 \tag{8.5}$$

Figure 8.7 presents a comparison of the statistical correlation to predict the trends from the work of Eichler et al. (2011) in a similar, confined geometry. (Please note that in this figure, instead of the scattered data, a trendline is drawn for illustrating the experiments of Eichler.)

As can be seen in the figure, the predicted values by the correlation (8.5) are consistently lower than the actual experimental values. However, given that the statistical model was developed on a different experimental setup (i.e., different tube sizes, turbulence levels, etc.) the

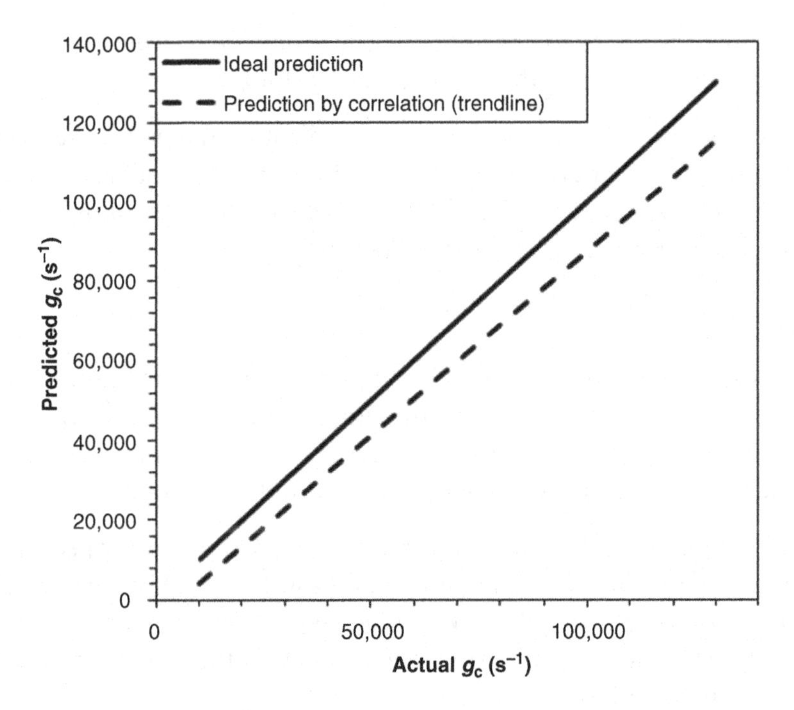

Fig. 8.7. Predicted g_C *values by the correlation versus the measured ones for confined tube.*

agreement can still be seen to be quite satisfactory. It should be noted that although the prediction of g_C by this correlation (Eq. (8.5)) can provide an estimate of flashback propensity, it is still only a rough estimate at best of the flashback propensity at practical gas turbine conditions because the effect of preheat temperature and pressure is not considered.

Experiments on the pressure dependence of the critical velocity gradient were performed by Fine (1958) for various fuel types, within a pressure range from subatmospheric to atmospheric at room temperature. Laminar and turbulent flows were investigated on water-cooled burner tube setup. The measurements led to the following approximate pressure dependence of the critical velocity gradient:

$$g_C \propto p^{1.35} \tag{8.6}$$

for H_2–air mixtures within the range $0.95 < \phi < 2.25$, for laminar and turbulent flows. However, this proportionality was found to become inaccurate for equivalence ratios considerably less than unity (no specific values were provided by the author). The results for g_C were observed to be independent of the tube diameter, except for $\phi > 2$.

Dependence of the critical velocity gradient on the preheat temperature was experimentally investigated by Fine (1959), at subatmospheric pressures. For laminar and turbulent H_2–air flames, the following proportionality was suggested by the experiments:

$$g_C \propto T^{1.52} \tag{8.7}$$

An alternative strategy for taking pressure and temperature effects into account may be the use of the proportionality (8.3). In this case, the scaling of the S_L^2 / α term with pressure and/or temperature can be calculated using detailed chemical kinetics, and taken over to scale g_C.

Mayer et al. (2011) and Sattelmayer et al. (2014) performed atmospheric experiments (preheat temperature 400 °C) on a hydrogen-fueled, swirl-stabilized burner (400 kWth) sketched in Figure 8.8 and observed different flashback types to occur depending on the changes in the geometry and operation conditions.

Swirl was generated by tangential entry of the main air through the four longitudinal slots arranged equidistantly around the circumference of the cone-shaped burner (Figure 8.8). A small fraction of air entered axially (unswirled) though the apex of the cone. The swirler was followed by a converging mixing section, and a short diffuser part. The latter could be dismounted. The fuel could be injected axially

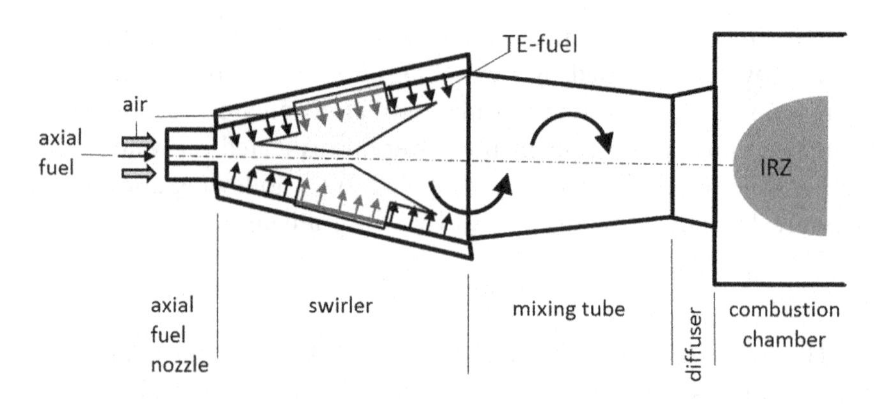

Fig. 8.8. Sketch of the burner and fuel injection concept.

through the apex of the conical swirler and/or through injection holes along the four trailing edges (TE), which were formed by the inner walls of the swirler slots and the inner conical surface (Figure 8.8). Injection through the TE holes, which will be considered here, was observed to exhibit almost the same flashback characteristics and NO_x emissions compared with perfect premixing. In the experiments, flashback was triggered by gradually increasing the equivalence ratio, while keeping the air flow rate constant.

Using the configuration with diffuser, the observed flashback type was the wall boundary layer flashback (WBLF). This can be explained by the decreased wall velocity gradients due to flow divergence in the diffuser, which increases the propensity for this type of flashback. The measured critical gradient values (Mayer et al., 2011) were observed to show a good agreement to the data of Khitrin et al. (1965) (displayed in Figure 8.4) obtained for unconfined tube burners. The critical velocity gradient concept assumes the equality of the flow velocity and laminar flame speed at the flame anchoring point in the boundary layer. This suggests the flame propagation speed to be the difference of both the quantities. In the experiments (Sattelmayer et al., 2014) the flame propagation speed in the axial direction was found to exhibit values around the laminar flame speed (5 m/s) and even higher, which would imply vanishingly small axial flow velocities. Sattelmayer et al. (2014) explained this by the computational and experimental findings of Eichler (2011), which revealed that the flame near the wall leads to streamline curvature and to the formation of a separation bubble. Thus, similar to CIVB-driven flashback (to be discussed in the next chapter), the flame propagates upstream in a zone with an almost zero axial velocity, which provides an explanation for the observed behavior.

In the configuration without a diffuser, the relatively high wall velocity gradients at the end of the converging section show a greater resistance against a wall boundary layer flashback. In this configuration, a CIVB-induced flashback was observed, first. After a certain upstream propagation of the cone/spherical-shaped flame front, the flame reaches the wall by turbulent flame propagation. At a position where the wall velocity gradient is no more sufficiently large (in the converging mixing section, the wall velocity gradient decreases in the upstream direction), the corresponding mechanism is activated and, thus, the CIVB-induced flashback was followed by the wall boundary flashback.

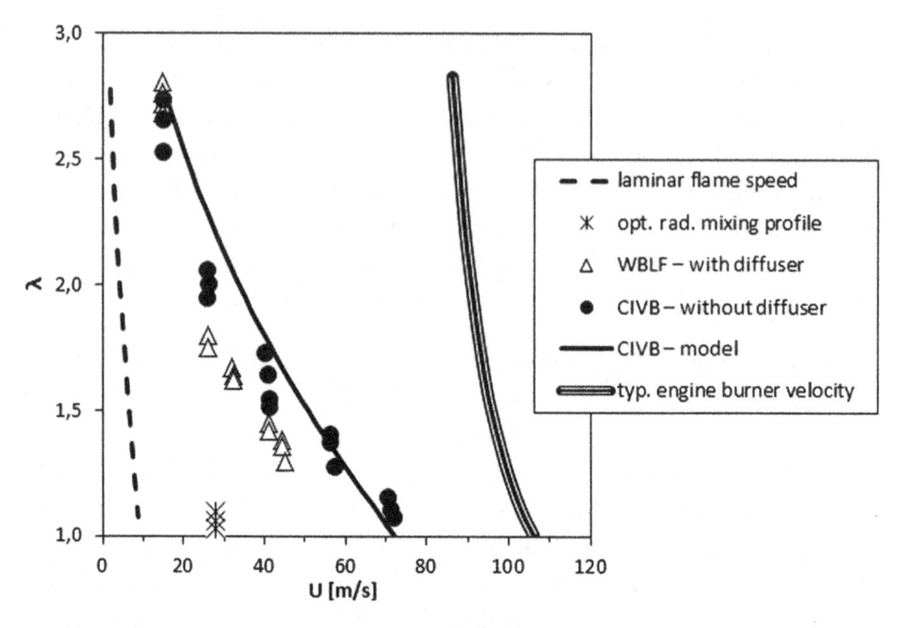

*Fig. 8.9. Comparison of flashback data for atmospheric hydrogen–air flames with preheat temperature 400 °C (*U*, bulk velocity).*

It was observed that the CIVB-induced flashback for the configuration without diffuser could be suppressed by allowing a portion of the air to enter axially through the apex of the cone (Figure 8.8), leading to larger axial velocities in the center. However, the simultaneous decrease of the near-wall velocities increased the wall boundary layer flashback propensity. Increasing the near-wall velocities at the expense of the center velocities would have a reverse effect and shift the dominating flashback mechanism from the wall boundary layer type to the CIVB type. Thus, one can conclude that the optimum is a well-balanced flow distribution with similar resistances against CIVB-driven and wall boundary layer flashback.

Figure 8.9 displays the measured (symbols) flashback limits for the configurations with and without the exit diffuser, which lead to wall boundary layer and CIVB flashback, respectively. For the CIVB flashback, the prediction by the Konle model, to be mentioned in the next chapter, is also displayed. Additionally the laminar flame speed is included, for comparison. The thick line on the right represents an estimate of the bulk flow velocity at engine conditions for a pressure drop of 2.5% over the burner.

The data shown in Figure 8.9 indicate a relatively good flashback resistance of the burner. For typical engine burner air excess ratios of 2, the approximated engine bulk flow velocities are nearly three times higher than the (atmospheric) flashback limit. Even at stoichiometric conditions the flashback limits remain substantially lower than the engine burner bulk velocity. The stars plotted on the lower edge of the figure (opt. rad. mixing profile) represent the flashback limits for a case with modified fuel and air injection to create lean conditions in the core and in the wall region, for increasing the resistance against both types of flashback. One can see that the flashback propensity can significantly be reduced further, by means of such measures. However, such measures may also tend to increase NO_x emissions.

The data presented in Figure 8.9 are for atmospheric pressure. Based on the pressure scaling (8.6) and the Blasius correlation (8.4), Sattelmayer et al. (2014) estimated that the prevention of wall boundary layer flashback at an engine pressure of 20 bar would require approximately 2.8 times higher velocity compared with the atmospheric data.

CHAPTER 9

Combustion-Induced Vortex Breakdown–Driven Flashback

Flows in gas turbine combustion chambers are almost exclusively swirled for flame stabilization (Chapter 2). The swirl number is normally so high that a so-called "vortex breakdown" (VB) occurs, i.e., recirculation zone on the burner axis is established [internal recirculation zone (IRZ)]. Speaking in terms of the time-averaged fields, along the burner axis, approaching the VB position, the axial velocity is sharply reduced to zero (stagnation point) within a small distance, and, further, to negative values in the IRZ, beyond the VB position. The flame is normally anchored at a position slightly upstream of the stagnation point, within the region of sharply decaying axial velocity, slightly ahead of the stagnation point where the local axial velocity and the turbulent flame speed are in balance. In swirling flames, a flashback in the core flow can occur, even when the turbulent flame speed is substantially smaller than the approach axial flow velocity. As opposed to flashback due to turbulent flame propagation, this mechanism is largely driven by the interaction of the heat release with swirling flow aerodynamics, leading to a transition of the VB characteristics (Fritz et al., 2004). This mechanism is quite commonly referred to as "combustion-induced vortex breakdown" (CIVB) (Fritz, 2003). Although this mechanism came into discussion rather in conjunction with swirlers without a centerbody, a similar effect is also observed in burners with a centerbody (Lieuwen et al., 2008b).

In discussing CIVB, the vorticity transport equation is useful, which is given as follows (Hasegawa and Noguchi, 1997):

$$\frac{D\vec{\omega}}{Dt} = \underbrace{(\vec{\omega}\cdot\vec{\nabla})\vec{u}}_{\text{I}} - \underbrace{\vec{\omega}(\vec{\nabla}\cdot\vec{u})}_{\text{II}} + \underbrace{\frac{1}{\rho^2}(\vec{\nabla}\rho\times\vec{\nabla}p)}_{\text{III}} + \underbrace{\nu\nabla^2\vec{\omega}}_{\text{IV}} \tag{9.1}$$

The right-hand-side terms of the equation are: I, vortex stretching; II, expansion; III, baroclinic torque; IV, viscous diffusion. Term "I" represents the enhancement of vorticity by stretching. Note that this term is zero for a two-dimensional flow. Term "II" represents the effect of expansion on the vorticity field, and becomes active only for variable density. In an expanding flow (e.g., due to combustion), the divergence of the velocity field is positive. This term would then lead to a decrease in the magnitude of vorticity. Term "III" becomes active only for variable-density flows, too, and leads to vorticity generation as a result of nonaligned density and pressure gradients. The term is zero if the density and pressure gradients are aligned, and attains its maximum value if they are perpendicular. Term "IV" describes the effect of diffusion on vorticity distribution. Note that the terms I, II, and IV become active if vorticity is nonzero. The baroclinic torque term "III" does not explicitly depend on vorticity, and appears as the only term that can produce vorticity in a vorticity-free flow field.

9.1 FLAME PROPAGATION IN VORTEX TUBES AND VORTEX RINGS

Experimental and theoretical studies on flame propagation on some generic types of vortical flows, such as unconfined vortex rings or tubes, have been performed since the 1970s. Early experiments in this area were performed by McCormack et al. (1972), where the burnable mixture was injected into the ambient air through a ring-shaped nozzle to form a vortex ring, which was then ignited in a plane perpendicular to the ring axis. After ignition, the flame front (FF) propagated symmetricly to the ignition plane, running along the vortex ring, around the axis of symmetry of the burner. A propagation speed U_f of approximately 15 m/s was measured, which was much higher than the expected flame speed. Theoretical studies on flame propagation in vortex tubes go back to Chomiak (1976), who postulated that the flame propagation is driven by the pressure increase (adverse pressure gradient) across the FF, caused by the volumetric expansion. As in many other theoretical studies that followed, a Rankine vortex was assumed (Chomiak, 1976). With given density and swirl velocity profiles, the radial pressure distribution was obtained by the so-called "radial equilibrium" condition, i.e., from $dp/dr = \rho w^2/r$. The sudden radial expansion of the vortex tube causes the swirl velocity to decrease in the core, which, in turn, implies, for the unconfined flow with imposed ambient pressure, a local increase of the pressure on the axis. This adverse pressure difference across the

FF can then cause the burned gas to get "sucked" into the upstream unburned gas. Chomiak (1976) estimated the mean pressure difference as $\Delta p \sim \rho_u W_{max}^2$ across the FF, where W_{max} denotes the maximum swirl velocity on the edge of the vortex core, leading to the following expression for the propagation speed (U_f):

$$U_f = C\sqrt{\sigma}W_{max} \tag{9.2}$$

where σ and C denote the unburned-to-burned density ratio and a model constant, respectively. Daneshyar and Hill (1987) performed a similar analysis and found the same relationship (Eq. (9.2)), where the constant C was calculated within the range $C = \sqrt{2/3} - \sqrt{2}$, whereas it was $C = 1$ in the derivation of Chomiak (1976). The investigations of Hasegawa et al. (2002) for unconfined, straight vortex tubes implied a rather linear dependence on σ. The following correlation was suggested for U_f:

$$U_f = 0.18(\sigma - 1)W_{max} + \sigma S_L \tag{9.3}$$

where U_f can be thought of as being composed of two parts, namely, a part due to vortex dynamics (the first right-hand-side term) and another part (the second right-hand-side term) representing flame propagation in a quiescent gas that also contains information on the laminar flame speed S_L.

Unlike the previously mentioned studies, Asato et al. (1997) observed a nonlinear dependence of U_f on W_{max}, where U_f was measured to be proportional to $(W_{max})^n$, with $n < 1$. They derived a more complex U_f expression compared with (9.2) and (9.3), incorporating additional information on the radius of the flame tip, which, in turn, was made dependent on W_{max}, and predicted the mentioned behavior qualitatively better compared with the Chomiak's model (Eq. (9.2)). Another model that relies on the pressure force for flame propagation is that of Ishizuka et al. (1998), which is more sophisticated than the previous models in that it not only considers the motion of the burned gas but also treats the flame as a reaction wave. The resulting expressions for U_f are rather complex (Ishizuka et al., 1998) and assume knowledge on the type of expansion (radial vs. axial) and on the radial extent of the flame compared with that of the unburned vortex core.

An alternative to the previously discussed mechanisms that commonly rely on the pressure difference as the driving agent, the "baroclinic push" mechanism was suggested by Ashurst (1996). Here, it is assumed that the flame propagation is induced by the azimuthal vorticity generated by interaction of pressure and density gradients. Assume an axisymmetric, inviscid vortex tube of a burnable mixture, with no gradients along the axis of rotation, with a given radial swirl (azimuthal) velocity profile $w = w(r)$ (a Burger's vortex was assumed by Ashurst, 1996), with a constant axial velocity component. Here, the vorticity vector has only one nonzero component, which is the axial vorticity created by the swirl velocity. The radially varying swirl velocity implies a positive radial pressure gradient via the aforementioned radial equilibrium. In this situation, if the vortex tube is ignited and a FF perpendicular to the vortex axis is created, the "baroclinic torque" term, which appears as a source term in the vorticity transport equation (Eq. (9.1)), becomes active and creates "azimuthal vorticity." Omitting all terms except the material derivative of vorticity and the baroclinic torque, the vorticity transport equation can be expressed as $D\vec{\omega} / Dt = (\vec{\nabla}\rho \times \vec{\nabla}p) / \rho^2$. One can see that in the case described, the generated azimuthal vorticity implies a rotation, which is aligned with the direction of the positive density gradient, i.e., in a direction pointing from the burned gas (low density) to unburned gas (high density), inducing a propagation of the flame into the unburned mixture. The resulting expression for U_f shows different qualitative characteristics to the previously discussed models as, for example U_f is proportional to the square of W_{max} and also dependent on the distance traveled by the FF. Numerical simulations of Ishizuka et al. (1998) confirmed an important role of the mechanism in the early stages of flame propagation.

Umemura and Takamori (2000) suggested another mechanism for the generation of azimuthal vorticity, which solely relies on the expansion of the vortex tube. According to this model, the vortex filaments are twisted by expansion of the burned gas in the radial direction, while the angular velocity must slow down for conserving angular momentum. Consequently, an azimuthal vorticity field is generated, which induces a propagation velocity toward the unburned gas. The mechanism is sketched in Figure 9.1.

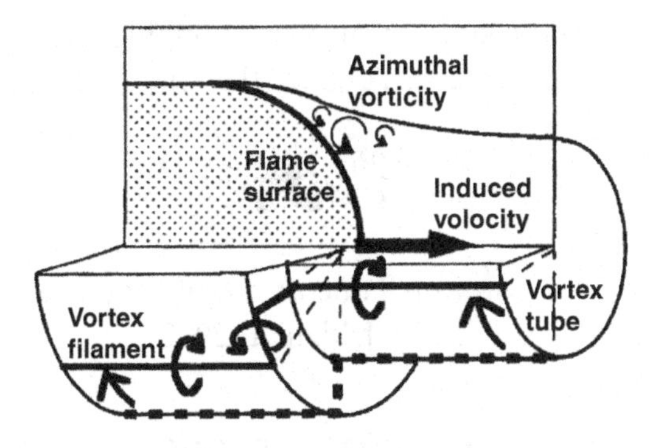

Fig. 9.1. Azimuthal vorticity generation (Umemura and Takamori, 2000).

The modeled U_f expression (Umemura and Tomita, 2001) reads as follows:

$$U_f = \sqrt{\frac{2\sigma + 1}{\sigma} W_{max}^2 + \sigma S_L^2} \tag{9.4}$$

The numerical simulations of Hasegawa et al. (2002) indicated that the baroclinic push mechanism plays a role in the initial acceleration of the flame in the vortex tube, and the azimuthal vortex generation mechanism due to vortex tube expansion (Umemura and Tomita, 2001) becomes dominant thereafter.

Takemura and Umemura (2002) extended the previous work (Umemura and Takamori, 2000; Umemura and Tomita, 2001) to the investigation on extinction conditions of a flame propagating in a vortex tube. They considered a combustible Rankine vortex immersed in an infinite domain of inert gas, which is ignited on its axis, in a region smaller than the vortex radius, resulting in two symmetric FFs propagating in opposite directions. In the numerical analysis, unsteady, compressible Navier–Stokes equations and scalar conservation equation were solved, assuming a single-step Arrhenius reaction and Lewis number of unity. The authors found that steady flame propagation is possible only when the maximum value of axial velocity induced at the vortex centerline, which depends on the heat input, is equal to or exceeds the propagation speed of azimuthal vorticity waves, which depends on the vortex strength. Otherwise, the flame gets elongated, becomes sharp at the tip, suffers cooling from the sides, eventually extinguishing. In that respect, Takemura and Umemura (2002) deduce the following stability limit:

$$\frac{\Omega \delta_L}{S_L} = 9.6 \tag{9.5}$$

A comprehensive review of flame propagation mechanism in vortex tubes and rings was provided by Ishizuka (2002).

9.2 CIVB IN GAS TURBINE COMBUSTORS

The flame propagation mechanisms outlined earlier essentially consider unconfined vortex tubes and rings. In gas turbine combustors, the swirling flow already exhibits a deliberately generated VB (and an IRZ), with an essentially constant position, where the flame is anchored and stabilized. The question is: how is the CIVB triggered in such a system?

Intensive investigations of flashback mechanisms in gas turbine swirl burners were performed by Sattelmayer and coworkers (Fritz, 2003; Kröner, 2003), where for the presently discussed flashback mechanism the designation CIVB was also coined. Figure 9.2 provides a sketch of the observed flow patterns during these CIVB-induced flashback experiments on a premix swirl burner without a centerbody (Kröner et al., 2007).

Figure 9.2a illustrates a stable flame with VB at the exit of the premixing tube. In the above-mentioned experiments, under unstable conditions leading to CIVB-driven flashback, a recirculation bubble was constricted

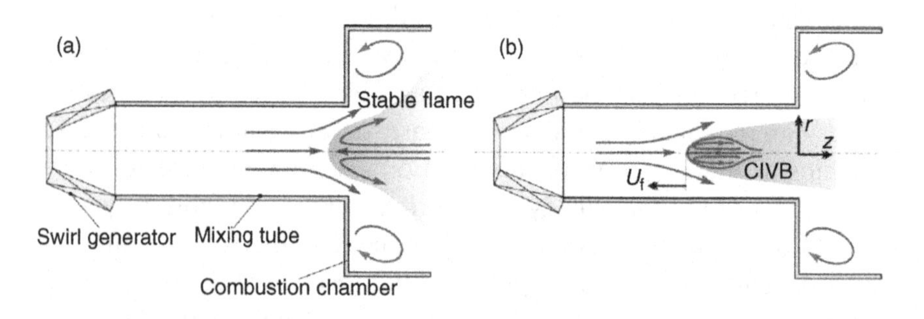

Fig. 9.2. Sketch of premixed swirl burner flow patterns for a (a) stable flame; (b) flame with flashback due to CIVB (Kröner et al., 2007).

from the burner tip as the FF was displaced upstream (Figure 9.2b). As the VB position and the attached FF moved upstream through the tube, the recirculation zone behind was closed within the tube, forming a recirculation bubble that also traveled upstream. The propagation continued until the flame reached the swirler and got stabilized there. In the experiments of Fritz (2003) on an atmospheric premixed 200 kW swirl burner without a centerbody, natural gas and hydrogen blends were used as fuel, for different modifications of the burner design. In general, starting from a stably burning lean flame, a CIVB flashback was induced by gradually increasing the equivalence ratio. This was achieved by increasing the fuel flow rate, while keeping the mean axial velocity and the swirl number nearly constant ($\sim 0.51 < S < \sim 0.54$). The CIVB flashback was observed when a critical value of the equivalence ratio was achieved ($\phi_{crit} \cong 0.75 - 0.8$). This role of the equivalence ratio on the onset of CIVB flashback may first be seen to indicate the role of kinetics, through the increased laminar flame speed. However, experimental investigations (Noble et al., 2006a, 2006b) show that the latter does not play a key role in the onset of CIVB-driven flashback. Instead, the increased expansion of the vortex tube was identified (Noble et al., 2006a, 2006b) as the cause, in agreement with the discussions in the preceding section.

Based on simplifying assumptions, such as an inviscid Rankine vortex flow and two unburned and burned fluids with different densities (constant in each fluid), Fritz (2003) developed an analytical expression for U_f for enclosed swirling flames exhibiting VB. Assuming that the turbulent flame speed is negligible against U_f, and constant total pressure across the FF, the following expression was derived ($\eta_1 = r_{vc,u}/R$, $\eta_2 = r_{vc,b}/R$, $r_{vc,u}$: vortex core radius upstream FF, $r_{vc,b}$: vortex core radius downstream FF, R: pipe radius):

$$U_f = W_{max}\sqrt{\frac{\eta_2^2(\eta_2^2-1)^2(2\eta_2^2\sigma-\eta_1^2(1+\sigma))}{\sigma(\eta_2^4(\eta_1^2-1)^2-\eta_1^4(\eta_2^2-1)^2\sigma)}} \tag{9.6}$$

In (9.6), η_2 is obtained by the conservation of axial momentum for the control volume. This leads to a rather complex equation for η_2 (Fritz, 2003), describing its dependence on η_1 and σ, which is nonlinear and can only be solved iteratively. It was found (Fritz, 2003), however, that Eq. (9.6) is useful for qualitative discussions, but not for quantitatively accurate results.

Noble et al. (2006a) investigated hydrogen addition effects on flashback on a test rig equipped with a premixed swirl burner with centerbody exhibiting VB, where reactant temperature and combustor pressure could be raised up to 500 K and 7 atm, respectively.

The authors defined two flashback mechanisms, with the following properties:

- Rapid flashback (discontinuous, abrupt): This "abrupt" flashback into the premixer was observed for rather high H_2 contents (60% or more) and/or high pressures. Comparison of the turbulent flame speed with the flow velocity was relevant for explaining this phenomenon. It was concluded that this flashback occurs through the boundary layers. This type of flashback was discussed in the previous section.
- Slow flashback (continuous, gradual): Here, a movement of the flame anchoring position monotonically upstream was observed, apparently due to a change in the location of VB, as the mixture fuel/air ratio increased. A qualitative illustration is provided in Figure 9.3.

Unlike the aforementioned "rapid" flashback, this "slow" flashback was observed to occur for rather low H_2 contents (less than 60%) and/or low pressures. In this regime, the hydrogen content of the fuel, and

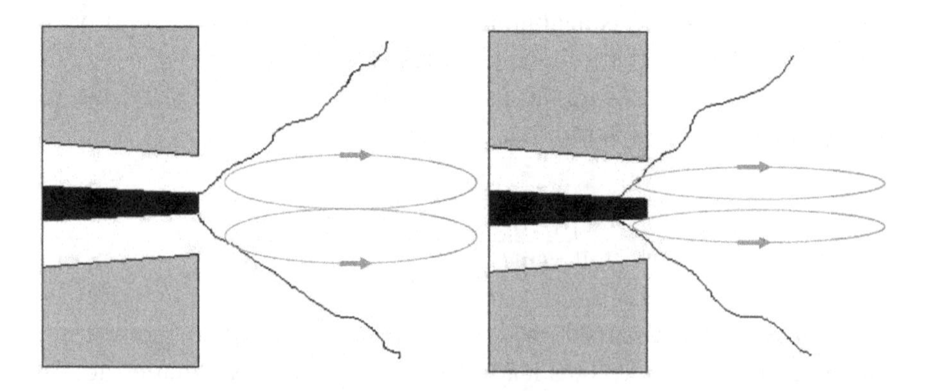

Fig. 9.3. Flame front and postulated recirculation zones for normal operation (left) and with flame propagated upstream ("slow" flashback) (right) (Noble et al., 2006a).

thus the flame speed, had far less effects on flashback characteristics; comparison of the turbulent flame speed with the flow velocity was not relevant for explaining this phenomenon, which was a clue for the predominant role of vortex aerodynamics in flame propagation. Thus, this type of flashback (Noble et al., 2006a) can be considered to belong to the category of CIVB.

Noble et al. (2006a) found that the temperature ratio (indirectly the pressure ratio) across the flame ($T_{burned}/T_{unburned}$) has the main effect and correlates with the flashback behavior. For obtaining an experimental evidence for the last item mentioned previously, Noble et al. (2006a) performed an experiment using a fuel composition, for which the maximum flame temperature and maximum flame speed occur at quite different equivalence ratios (ϕ). For the mixture 20% CH_4/20% H_2/60% CO, the maximum adiabatic flame temperature (T_{ad}) occurs at $\phi = 1.05$, whereas the maximum laminar flame speed (S_L) is encountered at $\phi = 1.24$. Variations of the S_L and T_{ad} with ϕ for this flame are shown in Figure 9.4.

The tests showed that as the equivalence ratio was swept from lean to rich, the flame moved farther into the premixer and occupied the farthest upstream point at the equivalence ratio corresponding to maximum flame temperature. Further increases in the equivalence ratio, corresponding to conditions where the flame speed was still increasing, but the flame temperature was decreasing, resulted in the flame moving back out of the premixer. This result seems clearly to show that upstream propagation of the flame closely correlates with the mixture's flame

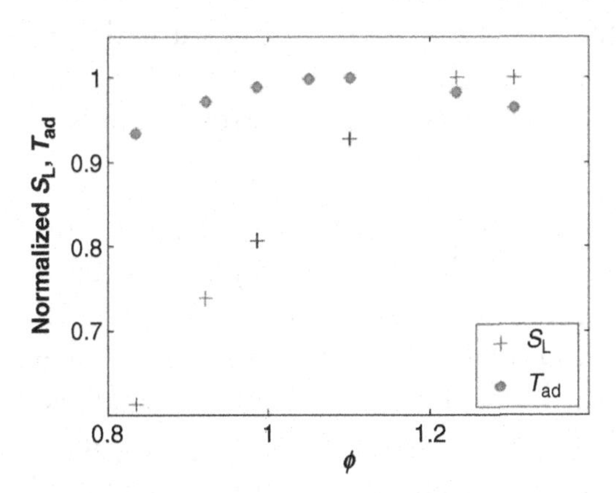

Fig. 9.4. S_L and T_{ad} versus Φ for a 20% CH_4/20% H_2/60% CO fuel mixture (Noble et al., 2006a).

temperature (i.e., the expansion ratio) but not its flame speed. This is in line with the earlier discussions in connection with the CIVB, where Noble et al (2006a) emphasize the role of the adverse pressure gradient, without however referring to other effects such as baroclinic torque or azimuthal vorticity generation by expansion. In a further study, Noble et al. (2006b) extracted a perturbation solution from the Darrieus–Landau flame stability analysis (Williams, 1985) for a flame with small sinusoidal wrinkles. Results indicated a monotonic growth of the pressure gradient ahead of the flame with increasing temperature ratio (T_b/T_u).

Lieuwen (2008) discusses the so-called "slow" flashback mentioned in Noble et al. (2006a, 2006b) and suggests an explanation: In many swirl burners, the fuel nozzle operates in a bistable regime (Brown and Lopez, 1990; Wang and Rusak, 1997) of swirling flows. At low swirl (LS) numbers, no VB is encountered. At high swirl (HS) numbers, VB occurs. However, at intermediate values, which are typical for practical systems (e.g., $S \sim 0.6–1.2$), the system has two possible states – no VB or VB. Lieuwen (2008) postulated that the so-called "slow" flashback occurs in this bistable region where practical designs operate. In this regime, the flow may be nominally axial, but could also, if appropriately perturbed, jump over the barrier and find the other energy functional minima corresponding to VB. The key proposal put forward by Lieuwen (2008) is that the flame, specifically, the adverse pressure gradient in front of the flame, can provide this perturbation to the flow, leading to the breakdown state in the premixing tube. It was discussed (Lieuwen, 2008) that the amplitude of this perturbation was proportional to two quantities, namely, the relative angle of the flame and the temperature ratio across the flow, as also discussed in Noble et al. (2006a, 2006b).

Kieswetter et al. (2007) considered an atmospheric, turbulent, premixed, methane–air flame generated by a swirl burner without centerbody in a geometrically axisymmetric test rig (Fritz, 2003). They performed a CFD analysis for analyzing the mechanisms leading to CIVB-driven flashback. In the unsteady, 3D analysis, turbulence was modeled by the LRR RSM, applied within the framework of a URANS formulation (Durbin and Reif, 2010). As the turbulent combustion model, the turbulent flame speed closure of Schmid et al. (1998) was used. The authors analyzed the predicted results in terms of the vorticity transport equation, for identifying the basic mechanisms that are responsible for CIVB-driven flashback. Their analysis indicated the baroclinic torque as the root cause of CIVB, as this term was observed to produce considerable levels of negative azimuthal vorticity in the vortex core. They also found that the essential features of CIVB were predictable applying a 2D URANS analysis, for the case considered. Additionally, Kieswetter et al. (2007) found that quantitatively accurate predictions of flashback limits could be obtained by the applied combustion model (Schmid et al., 1998).

The same flame (Fritz, 2003) was analyzed by Tangermann and Pfitzner (2009) using 2D URANS as well as 3D LES approaches. In URANS, an LRR RSM turbulence closure was used. The rate of the assumed single-step combustion reaction was modeled by the model of Schmid et al. (1998) as well as the model of Lindstedt and Vaos (1999). Wall quenching was modeled using two different approaches. The first model (Catlin and Lindstedt, 1991) simply assumes that no reaction can take place at temperatures lower than a presumed quenching temperature, setting the corresponding source term of the reaction progress variable locally to zero. The second model is the intermittent turbulent net flame stretch model of Meneveau and Poinsot (1991), which considers the influence of the turbulent strain rate and flame wrinkling on quenching.

In the LES, the dynamic Smagorinsky model (Sagaut, 2006) was used. The rate of the assumed single-step reaction was modeled following Mantel et al. (1996), along with an artificial thickening approach (Colin et al., 2000). Tangermann and Pfitzner (2009) found that the CIVB-driven flashback and its limits can be predicted sufficiently well by URANS and LES, however the LES results provided a better resolution of the flow structures and more insight for the initiation of CIVB. As gas turbines operate at high pressures, for evaluating the suitability of the approach for gas turbine applications, the applied combustion models were validated for high pressures, using the Bunsen flame experiments of Kobayashi et al. (1998). The Lindstedt–Vaos model was already shown by Brandl et al. (2005) to perform satisfactorily in that respect. In the calculations performed by Tangermann and Pfitzner (2009) for a range of pressures between atmospheric pressure and 10 bar, the accuracy of the

Schmid model was observed to deteriorate for pressures above atmospheric pressure. The model could be fitted for one operating pressure or a small pressure range but did not then, scale correctly to different pressures.

A more detailed analysis of an atmospheric, turbulent, premixed laboratory swirl flame was presented by Tangermann et al. (2010). In this study, along with the assumed single-step combustion, a BML-like (Libby and Williams, 1994) turbulent combustion model was used, solving a transport equation for the reaction progress variable. The source term, incorporating the undisturbed laminar flame speed and flame surface density, was closed using the correlations by Göttgens et al. (1992) and Boger et al. (1998), respectively. Different stages of the predicted flashback process are shown in Figure 9.5.

In the first stage, the flame propagates rapidly upstream from the combustion chamber to a position located slightly inside the mixing tube. In the early stage of the process, the movement of the recirculation zone is quite three-dimensional and unsteady, the position of the stagnation point being very volatile (Figure 9.5). When the CIVB stabilizes inside the mixing tube, a small recirculation bubble is formed and the remaining part of the recirculation zone is convected downstream. Subsequently, the actual flashback process starts. The VB and the flame tip propagate upstream at a rather constant velocity. During flashback, the stagnation point of the small recirculation zone is observed to be located always upstream of the flame tip. The propagation velocity was observed to be approximately 2.3 m/s in the experiment, which was predicted by LES with an almost perfect agreement (Tangermann et al., 2010). Note that the original, large IRZ in the combustion chamber is washed completely out at the final stage of flashback (Figure 9.5). The high axial velocities in the mixing tube (as a result of combustion and thermal expansion) drive the swirl number to such low values that a swirl number increase induced by the cross-section enlargement at the exit of the mixing tube no longer suffices to induce VB followed by an IRZ in the combustor.

Provided that the combustion model was tuned to reproduce one operating condition, by scaling the source term of the reaction progress variable equation, it was observed that the CIVB phenomena and the flashback limits could by predicted very accurately. Figure 9.6a shows

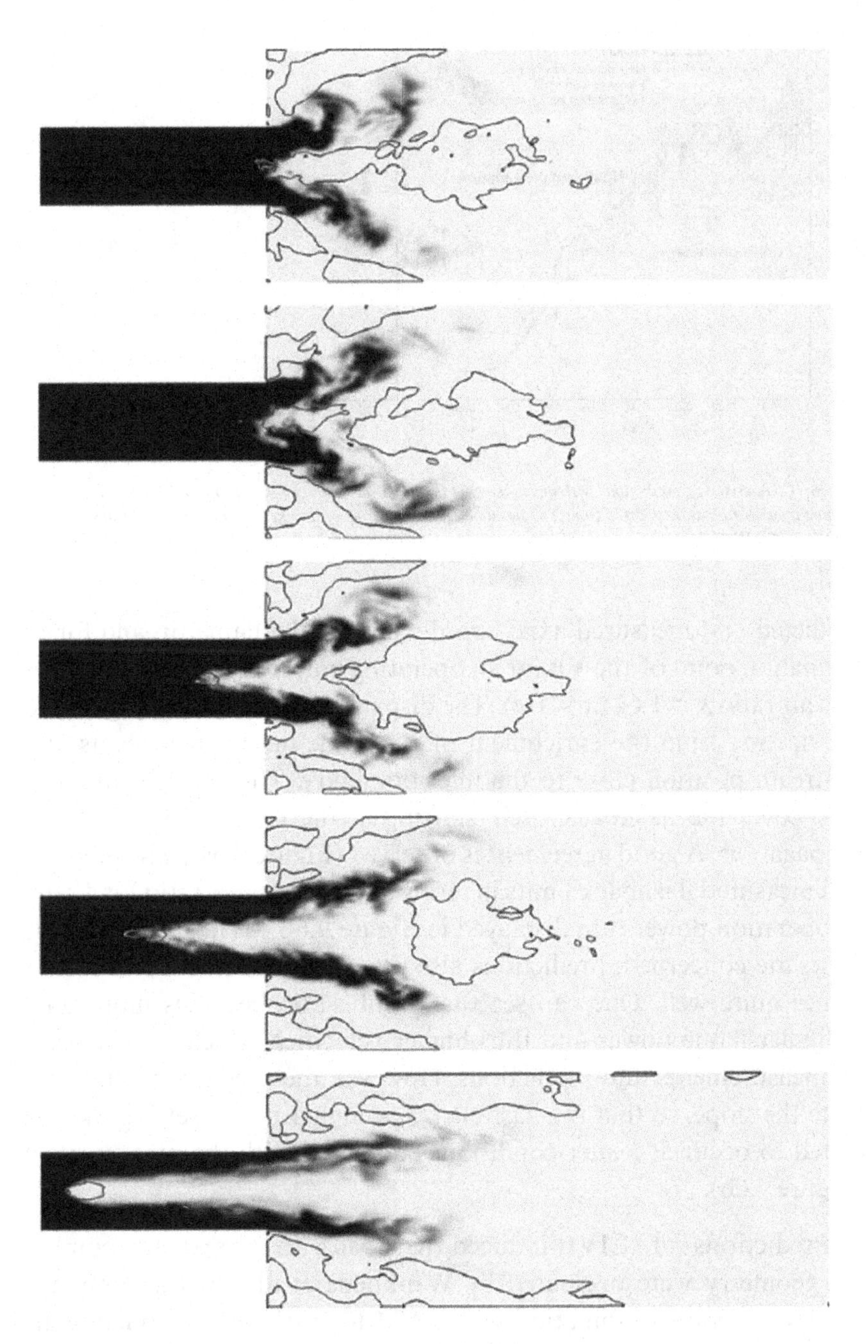

Fig. 9.5. Contour plots of the reaction progress variable and the isolines of zero axial velocity during CIVB-driven flashback. From top to bottom: formation of CIVB at 0.025 s after the final enrichment of the mixture, separation of the recirculation bubble at 0.03 s, and upstream propagation with a constant velocity at 0.045 and 0.06 s. Final stage in simulation stabilized by the inlet boundary at 0.075 s (Tangermann et al., 2010).

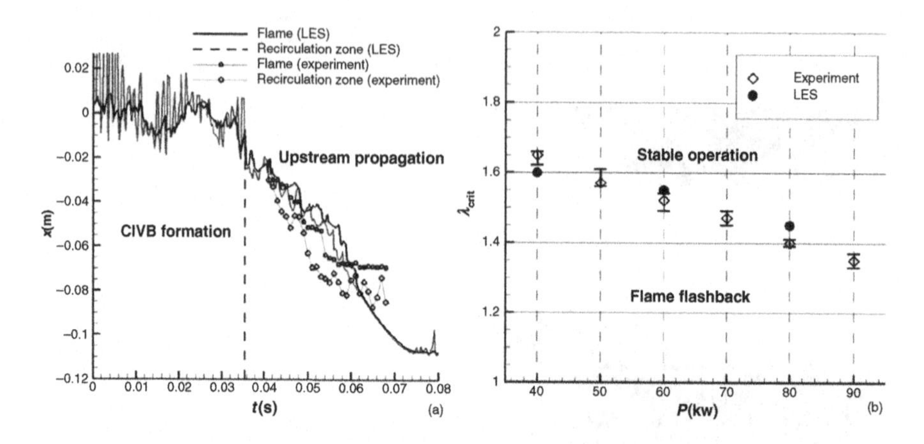

Fig. 9.6. Comparison of predictions with measurements for CIVB-driven flashback: (a) change of axial positions of flame tip and stagnation point in time; (b) flashback limits in terms of air excess ratio for an operating range from 40 to 90 kW (Tangermann et al., 2010).

predicted and measured axial coordinates for the flame tip and for the stagnation point of the VB for an operating power of $P = 40$ kW and excess air ratio $\lambda = 1.62$ ($\lambda = 1/\phi$). The displayed predictions in Figure 9.6a are starting from the enrichment of the flame until it reaches its most upstream position close to the inlet boundary. The experimental data only cover the flashback itself starting at the onset of the upstream propagation. A good agreement is observed (Figure 9.6a). The predicted and measured flashback limits in terms of the excess air ratio for a range of operation powers are displayed in Figure 9.6b. As far as the stability limits are concerned, predictions also seem to agree with the measured values quite well. One can see that combustion becomes more stable at higher flame power and thus higher velocities, which was indicated by measurements and predictions. However, the predictions underestimate the slope, so that the flashback at a higher power setting was predicted to occur at leaner conditions compared with the measurements (Figure 9.6b).

Predictions of CIVB-induced flashback in a realistic combustor geometry were presented by Wankhede et al. (2010), applying a 3D URANS in combination with RSM for turbulence modeling and Zimont's partially premixed turbulent combustion model. PVC and CIVB were predicted qualitatively, without, however, comparisons to measurements.

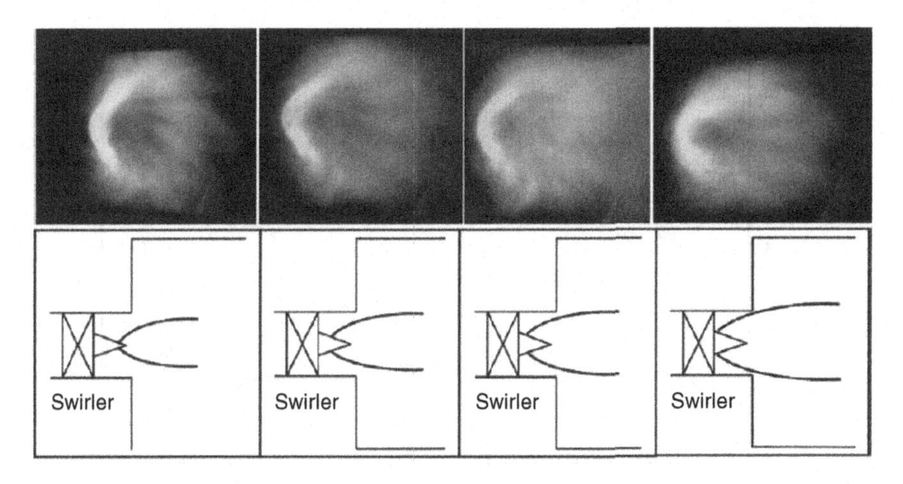

Fig. 9.7. Sequence of a CIVB-driven flashback phenomenon: (upper row) the photographic sequence; (lower row) the same qualitatively by approximate lines (Dam et al., 2011a).

Dam et al. (2011a) presented an experimental investigation of CIVB-driven flashback propensity for flames yielded from H_2–CO fuel blends and actual syngas mixtures. The effect of swirl number on flashback propensity was also discussed. Figure 9.7 shows a sequence of photographs and illustrations during the observed CIVB-driven flashback.

Figure 9.8 shows flashback limits for different fuel compositions and for two different swirl numbers.

For both swirl numbers, one can observe that even small concentrations of H_2 caused H_2–CO fuel mixtures to undergo CIVB-driven flashback at leaner conditions (Figure 9.9). In the previous sections,

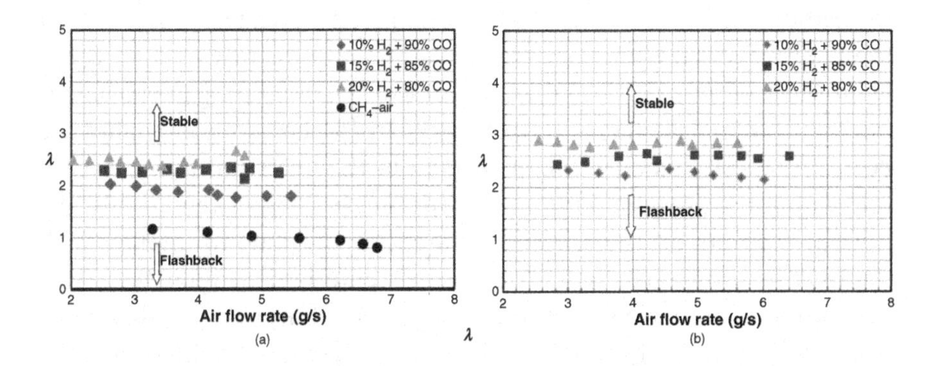

Fig. 9.8. Flashback limits for different fuel blends: (a) S = 0.97; (b) S = 0.71 (Dam et al., 2011a).

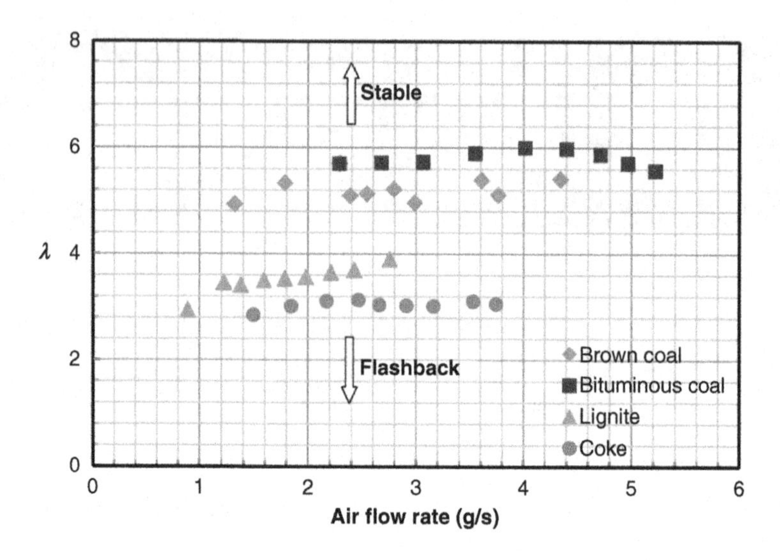

Fig. 9.9. Flashback limits for actual syngas fuels for S = 0.97 (Dam et al., 2011a).

the CIVB was identified as a mechanism that is basically driven by the burned-to-unburned temperature difference, without the kinetics playing a decisive role (Noble et al., 2006a). However, substitution of CO by H_2 does not necessarily cause a substantial change in the adiabatic flame temperature of the fuel blend. Dam et al. (2011a) explain the increased flashback propensity by hydrogen addition, by the changed kinetics and increased flame speed. However, according to the above-mentioned view (Noble et al., 2006a), the kinetics cannot be taken as the agent that triggers CIVB. However, the kinetics may be playing a role, "indirectly," in conjunction with the sustainability of the combustion reactions in the upstream flow. If the combustion reactions are locally quenched by the aerodynamic conditions, the flame cannot propagate upstream; even though the recirculation zone may penetrate into the mixing tube. Since hydrogen addition increases the resistance of the mixture against quenching, the observed flashback propensity by hydrogen addition might have been caused by this effect, if quenching in the upstream flow played a role without hydrogen addition. In swirl burners with a centerbody, the boundary layers on the centerbody immediately start to interact with the IRZ, as soon as the latter touches the centerbody (Figure 9.7). This makes it difficult to clearly distinguish the contributions by both mechanisms (CIVB-driven flashback vs. boundary layer flashback), so that the possible role of wall boundary layer flashback on the overall flashback

Table 9.1. Composition of Different Synthesized Gases Considered (Dam et al., 2011a)

Gasification	Type of Coal	CO (%)	H_2 (%)	CH_4 (%)	N_2 (%)	CO_2 (%)	Calorific Value (MJ/m^3)
Coal	Brown	16	25	5	40	14	6.28
	Bituminous	17.2	24.8	4.1	42.7	11	6.13
	Lignite	22	12	1	55	10	4.13
	Coke	29	15	3	50	3	6.08

behavior should also be considered. The results also show that the LS burner ($S = 0.71$) is more prone to CIVB flashback compared with the HS burner ($S = 0.91$) (Figure 9.8). An explanation to this behavior may, again, be provided by the quenching effects, which might have played a greater role in the case of higher swirl number.

Dam et al. (2011a) investigated the CIVB-driven flashback limits for actual syngas mixtures, the compositions of which are provided in Table 9.1.

CIVB flashback limits measured for the four different syngas compositions (Table 9.1) for $S = 0.97$ are shown in Figure 9.9. One can see the general trend that for a given air mass flow rate, the flashback behavior was, again, mainly dominated by the H_2 percentage in the fuel blend.

A rather recent CFD investigation on CIVB-driven flashback was provided by De and Acharya (2012). An unconfined laboratory swirl burner with a centerbody was numerically investigated, where the results were also compared with experiments. Different methane–hydrogen fuel blends were used, while keeping the equivalence ratio constant at 0.7. The turbulence was modeled by LES, adopting a dynamic Smagorinsky subgrid-scale model, along with a thickened flame approach for combustion modeling. Methane combustion reaction was modeled by a two-step chemistry, incorporating CO as the intermediate species, where hydrogen combustion was modeled by a single-step reaction. The reaction rates were modeled by Arrhenius expressions. A fairly good agreement between the predictions and measurements was observed, as far as the time-averaged velocity field is concerned. The size (length and width) and strength of the central recirculation bubble were observed to decrease with the increase in hydrogen content as shown in Figure 9.10.

This was explained (De and Acharya, 2012) by the higher combustibility of hydrogen resulting in increased temperatures in the FF. The higher temperatures (thus, lower densities) led to higher axial velocities

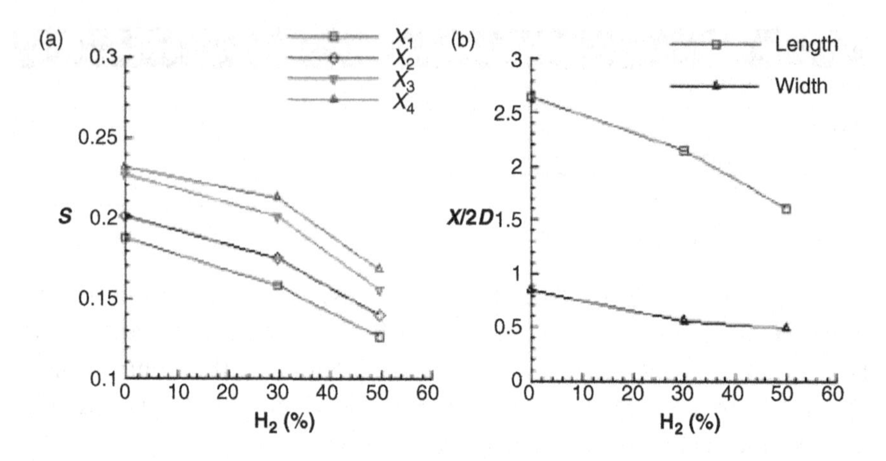

Fig. 9.10. Effect of hydrogen mole fraction in methane–hydrogen fuel blend on (a) swirl number [X_1, X_2, X_3, X_4 = (X/2D): 0.40, 0.79. 1.18, 1.58] and (b) recirculation bubble size (D: centerbody diameter, X: axial coordinate measured from burner exit) (De and Acharya, 2012).

($u \sim 1/\rho$, due to continuity), while the swirl velocity did not change significantly ($w \sim 1/\rho^{1/2}$, due to radial equilibrium). Thus, the swirl number decreased with increasing hydrogen fraction, which was, consequently, accompanied by a reduction of the size of the recirculation bubble (Figure 9.10).

Figure 9.11 shows the time-averaged temperature profiles at different hydrogen percentages in the fuel. Although the FF reaches into the mixing tube for pure methane combustion, it moves to further upstream positions with increasing hydrogen content of the fuel, moving all the way to the fuel injection point and getting stabilized in its wake.

De and Acharya (2012) analyzed the results in terms of the vorticity transport equation (Eq. (9.1)) and found that the stretching and baroclinic torque terms produced negative azimuthal vorticity to provoke CIVB-driven flashback, while the expansion and dissipation terms were practically cancelling out.

A recent computational investigation of CIVB-induced flashback was presented by Tian et al. (2014). The assumed fuel was hydrogen diluted with nitrogen (60% H_2 + 40% N_2). In this study, as in the previous study (De and Acharya, 2012), a swirler with centerbody was considered, while the majority of the previous investigations on CIVB considered swirlers without centerbody. There are differences in the

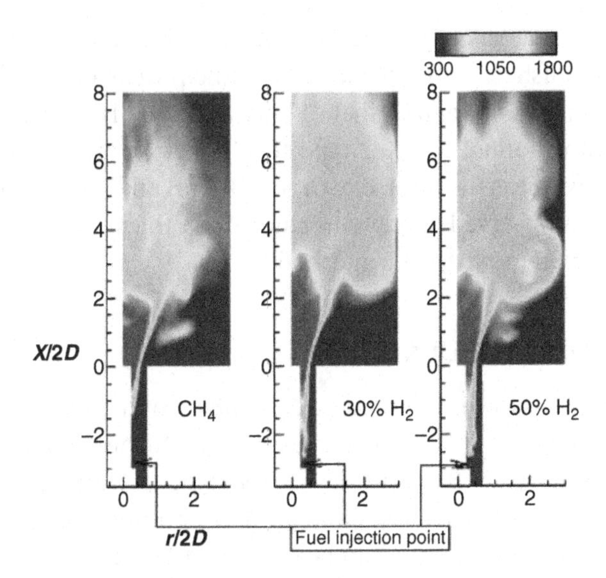

Fig. 9.11. Time-averaged temperature contours (K) in the near field of burner outlet for different fuel compositions (D: centerbody diameter, X: axial coordinate measured from burner exit (De and Acharya, 2012).

flow characteristics of both configurations. Principally, without centerbody, harmful effects of flashback are observed after the flame has reached the swirler. With centerbody, the flame damages the hardware as soon as it enters the premixer. Second, without centerbody, the flame propagates against the maximal axial velocity in the central parts of the burner, which is not the case with centerbody. A 2D RANS analysis was performed, based on RSM. The combustion model based on a single-step global reaction was the so-called finite-rate/eddy-dissipation model, where the time-averaged reaction rate was taken as the minimum of these two rates. In addition, the resulting reaction rate was set to zero if the local mean temperature was below an assumed ignition temperature (800 K).

In the investigation of Tian et al. (2014), for $\phi = 0.34$, the flame and the IRZ were stabilized in the combustion chamber (no flashback). With an increase of ϕ to 0.50 a low-velocity region was formed along the centerbody surface, which paved the way for upstream flame propagation. This took place in the wall boundary layer. For $\phi = 0.52$, the FF moved further upstream, where a thickening of the "head" of the negative velocity region was observed, while the flashback mode was still wall boundary layer flashback. A further increase of ϕ to 0.53 promoted CIVB flashback. The upstream negative velocity region was enclosed and a new recirculation zone was established. Subsequently, the flame propagated to this recirculation zone and was stabilized there. Thus, a shift in flashback mechanism changing from boundary layer type to CIVB type was observed (Tian et al., 2014), whereas the reverse path (first CIVB and then boundary layer flashback) was observed in previous studies (Baumgartner and Sattelmayer, 2013) on swirlers without centerbody.

9.3 THE ROLE OF QUENCHING

As already discussed in the previous chapters, an increase of the turbulence intensity leads, first, to an increase in the turbulent flame speed up to a maximum value. By a further increase beyond this point, the turbulent flame speed starts to decline until the point of flame quenching. The quenching phenomenon can be explained by a comparison between the turbulent mixing and chemical timescales. At the quenching point, the chemical reaction rate is no more sufficient to deliver sufficient heat, within the lifetime of an eddy, which is sufficient to ignite the unburned mixture. In case of CIVB, an upstream propagation of the flame in the mixing tube can be suppressed by quenching effects in the mixing tube, although the VB, i.e., the IRZ, might have propagated upstream. Thus, quenching effects in the mixing tube can guard against flashback. Therefore, although the basic mechanisms that trigger CIVB are not kinetically influenced, the kinetics play a role through quenching of the flame as it propagates upstream.

Figure 9.12 shows the experimentally obtained (Kröner et al., 2007) CIVB-driven flashback limits for atmospheric laboratory flames using a swirler without centerbody. In the figure, the measured flashback limits are displayed in terms of excess air ratio (λ_{cr}: critical excess air ratio) as a function of air mass flow rate and preheat temperature for two different fuel compositions.

In general, one can see that the flashback limit moves toward richer mixtures with increasing air mass flow rate, which is a plausible trend. Nevertheless, one can also see that the rate of change of λ_{cr} with air mass flow rate is rather weak, implying that the flashback propensity may not substantially be improved by adjusting burner mass flow rate, in practical applications.

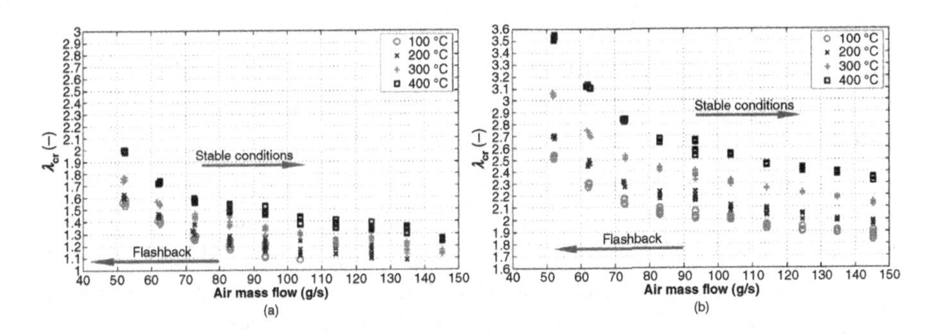

Fig. 9.12. CIVB-driven flashback limits (λ_{cr}) as a function of air mass flow rate and preheat temperature for different fuels: (a) 100% CH_4; (b) 85% CH_4 + 15% H_2 (mass fractions) (Kröner et al., 2007).

In Figure 9.12, one can also observe that the flashback limits move toward leaner conditions with increasing preheat temperatures. Since higher preheat temperatures imply higher velocities, and considering the fact that initiation of CIVB is not kinetically controlled, one would rather expect the contrary, i.e., a decline in the flashback propensity with increasing preheat temperature. However, the observed trend, (Figure 9.12), was explained (Kröner et al., 2007) by the increase of chemical reactivity, i.e., by the increase of quenching resistance of the unburned mixture at high preheat temperatures. Similarly, hydrogen addition (Figure 9.12) causes a further increase of the flashback limits toward even leaner mixtures through the same effect. Thus, for CIVB-driven flashback to occur, two conditions need to be satisfied: (1) the propagation speed U_f (as a result of VB aerodynamics) must exceed the axial flow velocity on the mixing tube axis; (2) the conditions in the mixing tube must be such that combustion reactions in the propagating FF are not quenched.

Kröner et al. (2007) provided an analysis of timescales to achieve a correlation for the quenching effects during CIVB-driven flashback. They proposed a description through a "quenching constant" C_{quench}, defined as follows:

$$C_{quench} = \frac{\tau_c}{\tau_u} \tag{9.7}$$

where τ_c and τ_u represent chemical and flow timescales, respectively. A more traditional/theoretical approach would be to attempt a description through the Damköhler or the Karlowitz number. Kröner et al. (2007) preferred a description through C_{quench}, since in the classical definitions of Damköhler and Karlowitz numbers, the flow timescale is estimated through turbulence velocity and length scales, which are rather difficult to evaluate in technical systems. The flow timescale is estimated as $\tau_u = D/U$, where D and U denote the mixing tube diameter and mass mean axial velocity, respectively. A possibility for the estimation of τ_c would be to calculate it from the equation $\tau_c = \delta_L/S_L$, which is frequently used for similar purposes. However, Kröner et al. (2007) do not find this approach to be convenient, since it assumes complete combustion, whereas the quenching process is inherently related with incomplete combustion. Thus, the chemical timescale is obtained from zero-dimensional PSR calculations based on detailed reaction mechanisms, i.e., $\tau_c = \tau_{\text{PSR}}$ (τ_{PSR}: timescale for the extinction of PSR). As already discussed in the previous chapters, preferential diffusion plays a role in the combustion of hydrogen blend fuels. For taking this effect into account, Kröner et al. (2007) finally suggest the following definition of a quenching constant C^*_{quench}:

$$C^*_{\text{quench}} = \text{Le}_F \cdot \tau_{\text{PSR}} \cdot \frac{U}{D} \tag{9.8}$$

where Le_F denotes an average Lewis number of the fuel mixture. Figure 9.13 displays the calculated C^*_{quench} values for different fuels for a burner configuration considered in Kröner et al. (2007).

One can see (Figure 9.13) that the quenching constant takes values around 0.03 for all experiments ($C^*_{\text{quench}} \sim 0.03$), for sufficiently large Reynolds numbers. Kröner et al. (2007) report, however, that this (approximate) constant number is not universal but can take different values for different burner configurations. However, this is still a notable result. Its technical significance stems from the fact that once C^*_{quench} has been determined for a specific burner on basis of a few experiments, the flashback limits can be predicted for a wide range of operating conditions and fuels, as argued by Kröner et al. (2007). At the same time, a pressure scaling of the experimentally determined flashback limits using a constant quenching constant without validation at the pressure of interest is not recommended due to the nonmonotonic behavior

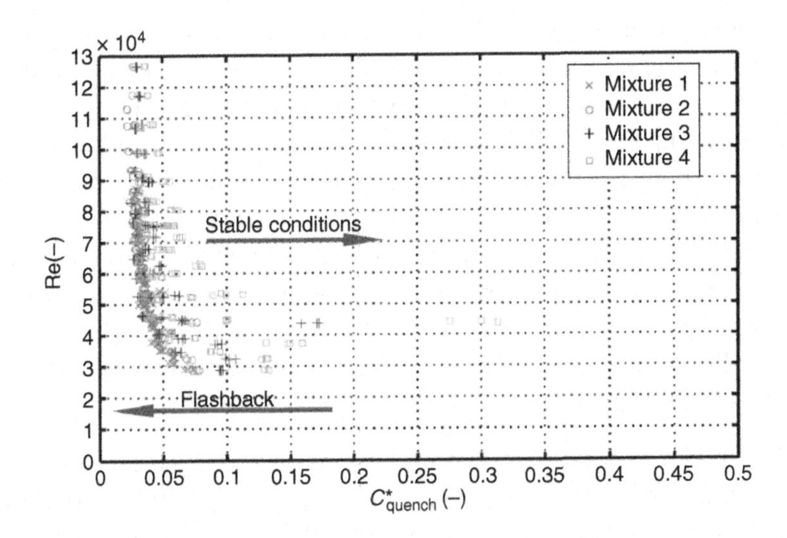

Fig. 9.13. Quenching constant for a burner configuration (BC1) at the flashback limit for different CH_4/H_2 fuel mixtures (H_2 mass fraction in the mixture – mixture 1: 0; mixture 2: 5%; mixture 3: 10%; mixture 4: 15%) (Kröner et al., 2007).

of the predicted pressure influence on the chemical timescale (Kröner et al., 2007), as pressure also affects the turbulence structure.

For the same purpose, Dam et al. (2011a) follow a different approach, i.e., the so-called Peclet number approach, which was previously used to describe lean blowout (Strakey et al., 2007), assuming a similarity between the processes. Here, a proportionality between the square of the "flow Peclet number" ($Pe_U = UD/\alpha$) and the "flame Peclet number" ($Pe_{SL} = S_L D/\alpha$) is assumed, where an equality is established, introducing a proportionality constant C (quench constant) as follows:

$$\frac{UD}{\alpha} = C\left(\frac{S_L D}{\alpha}\right)^2 \tag{9.9}$$

Dam et al. (2011a) presented experimental correlations between the flow Peclet number Pe_U and the flame Peclet number Pe_{SL} for different fuel compositions (10% H_2 + 90% CO and 20% H_2 + 80% CO) obtained using two different swirl burners (with S = 0.97 and 0.71). One could see that at both swirl numbers, the fuel-containing lower fraction of H_2 yielded the higher quench factor and both swirlers yielded a similar value of the quench constant for similar fuel compositions. For the 10% H_2 + 90% CO fuel composition the value of C was found to be approximately 0.36 and for the 20% H_2 + 80% CO fuel composition the value was approximately 0.23. Therefore, the quench constant C was dominated mostly by the fuel composition rather than the swirl strength. The quench constant C (Eq. (9.9)) by Dam et al. (2011a) has a certain similarity to C^*_{quench} (Eq. (9.8)) introduced by Kröner et al. (2007). However, C^*_{quench} (Eq. (9.8)) does not show a substantial dependence on hydrogen content of the fuel, whereas C (Eq. (9.9)) does. This may be due to the assumed explicit dependence of C^*_{quench} on the mixture Lewis number, whereas such a dependence was not formulated for C.

The model of Kröner et al. (2007) based on C^*_{quench} (Eq. (9.8)), which performed rather well for high Reynolds numbers, i.e., high burner mass flow rates, was improved by Konle and Sattelmayer (2010). They found that for burners operating at lower Reynolds numbers, where combustion takes place rather in the corrugated flamelet regime (Chapter 2), the chemical timescale based on a PSR model, as used by Kröner et al. (2007), does no more deliver sufficiently accurate results. According to the model of Konle and Sattelmayer (2010), in this regime, sufficient baroclinic torque for CIVB is produced, when the distance between the VB stagnation point and the FF undershoots a critical value, Δx_{crit}. Then, another chemical timescale, τ_b, was suggested, which was assumed to be proportional to $\Delta x_{crit}/S_L$. Assuming, further that Δx_{crit} is proportional to burner diameter D, the new timescale was estimated by $\tau_b \sim D/S_L$. Combined with the flow timescale $\tau_u = D/U$, this led to the following simple expression for the characteristic constant C_b:

$$C_b = \frac{U}{S_L} \tag{9.10}$$

simply stating that the ratio of the bulk velocity and the laminar flame speed should be constant for all operations near the onset of CIVB. Thus, according to the model, the experimental determination of this ratio via at least one reference measurement allows the prediction of CIVB-driven flame flashback for the entire operation range.

The experimentally obtained CIVB limit for the so-called TD1 swirl burner (without centerbody) of Konle and Sattelmayer (2010), as well as correlation results, is plotted in Figure 9.14a, for $d = 12$ mm, where d is the diameter of the rather tiny jet on the centerline of the burner that introduces unswirled core flow. Figure 9.14b displays the same for two different values of d ($d = 9$ and 15 mm).

For the geometry with $d = 12$ mm (Figure 9.14a) a characteristic constant $C_b = 0.412$ according to Eq. (9.10) was determined using the critical operation point $P_{th} = 60$ kW and $\lambda_{crit} = 1.52$ as the reference. Using this value of the constant, the critical values of λ were then predicted for the entire operation range (Figure 9.14a) of the investigated burner by calculating the corresponding timescales (the solid line), which is in good agreement with the experimental data (symbols).

Variations in the burner configuration, such as the changes in the axial inlet diameter d, cause a variation in the core radius of the swirl flow and lead to different values of the characteristic constant C_b (Figure 9.14b), confirming that the latter is burner specific and needs to be determined for a certain burner type by at least one reference measurement. Again using the conditions at $P_{th} = 60$ kW as basis, the estimated constant C_b was 0.452 for $d = 9$ mm and 0.352 for $d = 15$ mm (Figure 9.14b). The lower value for the larger diameter indicates better flashback resistance. One can see that the correlation law of Konle and Sattelmayer (2010) holds quite well for the considered variations of the burner configurations, the maximum deviation between the predicted and measured λ_{crit} being less than 0.025

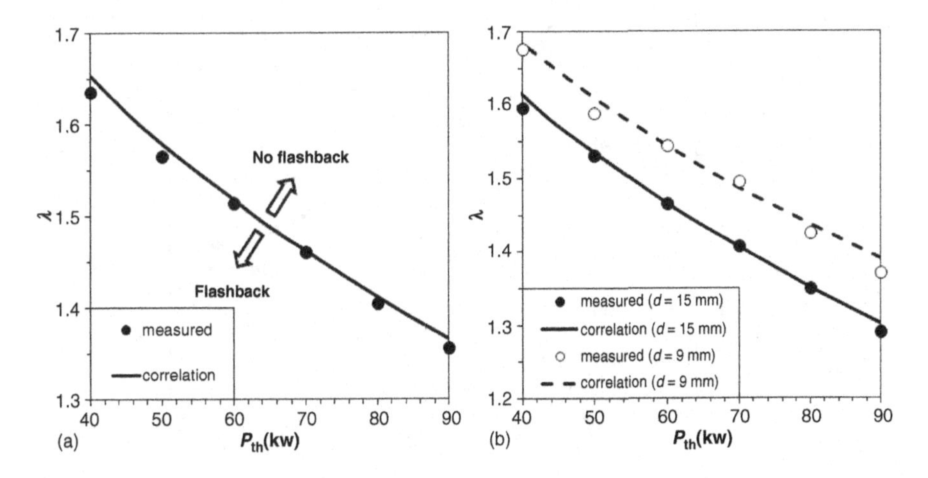

Fig. 9.14. Experimental CIVB limits and correlation (Eq. (9.10)) with $C_b = 0.412$) for (a) d = 12 mm; (b) d = 9 and 15 mm.

(Figure 9.14). Konle and Sattelmayer (2010) tested the proposed correlation with respect to its applicability for the prediction of CIVB limits for preheated mass flows as well. They reported that the model delivers sufficiently accurate results for changes in the preheat temperature, without needing an adaptation of the estimated constant C_b.

For coverage of a broader range of burner operation conditions, Konle and Sattelmayer (2010) suggest a combined use of correlations based on τ_b (Eq. (9.10)) and τ_{PSR} (Eq. (9.8)), depending on the burner operation range. This is demonstrated in Figure 9.15, for the burner investigated by Kröner et al. (2007).

The critical operation point, P_{th} = 125 kW and λ_{crit} = 1.44, was the reference point used, which resulted in the following values of the geometry-specific constants: $C_b = 0.153$ and $C_{quench} = 0.030$. The two curves show the correlation laws according to Eqs. (9.10) and (9.8), respectively. For stable operation, the burner needs to be operated above the critical air excess ratios predicted by both correlation laws. The validity for Eqs. (9.8) and (9.10) ends at the intersection of the curves: for low to

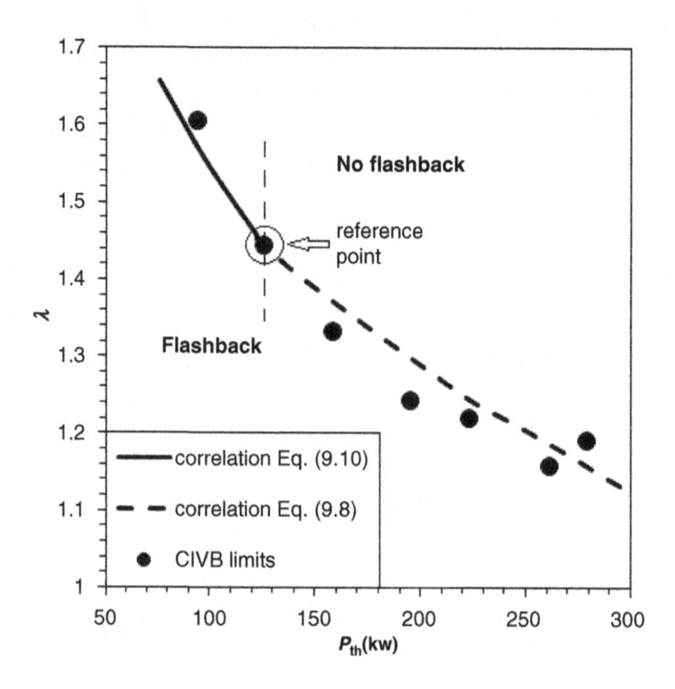

Fig. 9.15. Experimental CIVB limits and correlations with the coupled correlation laws.

moderate mass flow rates, the assumption of corrugated flames is obviously also valid for the burner investigated by Kröner et al. (2007) and predicts the limits for the onset of CIVB quite well. With increasing mass flow rates the model of Kröner et al. (2007), i.e., Eq. (9.8), is better suited for the prediction. The combined use of both timescales allows a good prediction over the entire operation range of the burner, the maximum deviation between the modeled and measured λ_{crit}, again, being smaller than 0.025.

Cheng et al. (2009) performed flashback experiments on a low-swirl burner, called "low-swirl injector" (LSI), which does not exhibit VB. Hydrogen–methane blends were used as fuel, with varying hydrogen fraction up to 100%. The authors observed that the critical equivalence ratio, where flashback has occurred, increased with bulk flow velocity and decreased with hydrogen content of the fuel. The prevailing flashback mechanism was not discussed in detail (Cheng et al., 2009). An interesting result of the study was that the LSI showed a higher flashback resistance compared with a high-swirl burner (SimVal 30° Swirler; Sidwell et al., 2006) exhibiting VB. Following the approach of Kröner et al. (2007), Cheng et al. (2009) calculated a quenching constant C_{quench} (Eq. (9.7)) for quantifying the difference in the flashback behavior of the LS and HS burners. In difference to Kröner et al. (2007), Cheng et al. (2009) calculated the chemical timescale from $\tau_c = \delta_L/S_L$, and did not consider a Lewis number correction (Eq. (9.8)). In Figure 9.16, the determined quenching constant C_{quench} values are shown as a function of the density ratio of unburned to burned gases ρ_u/ρ_b, for both burners.

Flashback resistance is associated with small values of C_{quench} that represents a short chemical time at flashback compared with the mean convective time. As seen in Figure 9.16, the C_{quench} values of LSI are generally lower than those from the HS burner, showing that the LSI has a lesser propensity to flashback.

In swirl burners, CIVB is, of course, not the only flashback mechanism in the core flow. Depending on flow conditions, flashback due to flame propagation can also occur. Blesinger et al. (2010) investigated both forms of flashback on the variations of the same (flashback due to flame propagation in the core flow was termed "turbulent burning

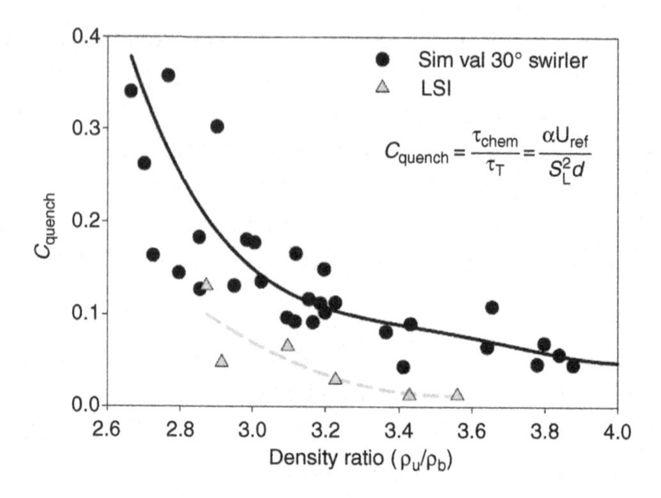

Fig. 9.16. Comparison of the flashback criterion C_{quench} of low-swirl (LSI) and high-swirl burners for CH_4–H_2 blend fuels with 60–100% H_2, for 0.101–0.810 MPa, 500–600 K, and $0.4 < \phi < 0.7$ (Cheng et al., 2009).

along the vortex axis" (TBVA)) laboratory burner, exhibiting LS and HS. The flashback was triggered by increasing the equivalence ratio, at a critical value of which flashback occurred (with different critical values for the two forms of flashback). For the LS case, under stable conditions, the VB was located just downstream the mixing tube exit. At the critical equivalence ratio, the CIVB-driven flashback was observed. For the HS case, the VB was located at the upstream end of the mixing tube under isothermal conditions. In the reacting case, the flashback is then caused by turbulent burning in the inner recirculation zone (IRZ) that extends upstream along the mixing tube. For $\phi < \phi_{crit}$ this type of flashback is prevented by turbulent flame quenching inside the part of the recirculation zone that extends into the mixing tube. For $\phi \geq \phi_{crit}$ the combustion is sufficiently strong to compensate for the heat losses and to stabilize inside the recirculation zone upstream of the mixing tube exit (Figure 9.17). Flow conditions for TBVA prevail in the HS burner configuration. As discussed previously, similar to TBVA, turbulent flame quenching inside the recirculation zone is also important for CIVB and may stop flame propagation.

As shown in Figure 9.18, the aerodynamics of both flashback types can be observed on the basis of simultaneous recordings of the axial velocity field (left: planar PIV) and the flame position (right: flame

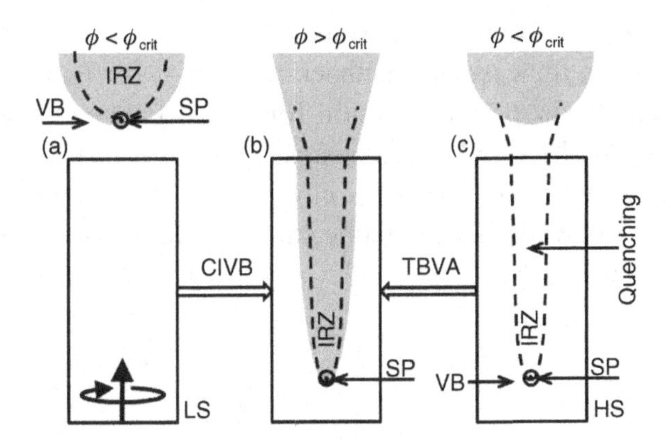

Fig. 9.17. Qualitative illustration of flame (shaded area), mixing tube and recirculation zone (dashed line) for different operation conditions (SP: stagnation point): (a) stable flame for low swirl level; (b) flame after flashback (similar for flashback caused by CIVB and flashback caused by TBVA); (c) stable flame for high swirl level.

luminescence). ϕ and Re are set to values at the onset of flashback. Under these conditions the flame oscillates upstream and downstream in time.

In the case of CIVB-driven flashback, the flow field is strongly altered by the flame. During the upstream flame propagation, the FF is

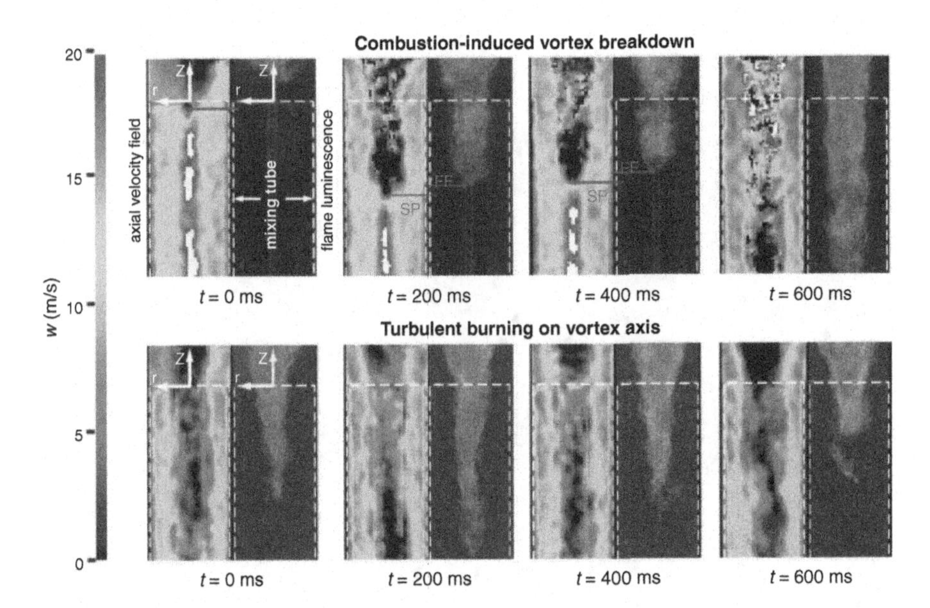

Fig. 9.18. Axial velocity field (2D) and flame luminescence during flashback due to CIVB (LS, Re = 28,000, ϕ = 0.66) and TBVA (HS, Re = 28,000, ϕ = 0.63) (Blesinger et al., 2010).

attached to the VB position [marked by the stagnation point (SP) in Figure 9.17)]. In the case of flashback by TBVA, the flame propagates into an already existing recirculation zone. No change of the flow field is necessary. At the onset of flashback caused by TBVA the combustion inside the recirculation zone is close to extinction. This is apparent in Figure 9.18 considering the flame entering and leaving the turbulent recirculation zone.

CHAPTER 10

Flameholding by Fuel Injection Jets

In the normal operation of a premix burner, the fuel injection jets are not exposed to the flame front before they are mixed out, and they do not create a cause for flashback. However, once a flashback has occurred, and the flame front has reached the injection holes, the flame can anchor in the near-field of the injection jets, even after the cause for the flashback is over, since highly inhomogeneous flow and mixing field comprising recirculation zones provide favorable conditions for flame stabilization. For a safe operation, in spite of any disturbance that may temporarily occur, this possibility needs to be precluded. This means that the flow, turbulence, and mixing fields in these regions must be designed in such a way that local extinction processes prohibit any flame stabilization.

10.1 JET IN CROSS-FLOW

The generic flow configuration for fuel injection jets is the so-called "jet in cross-flow." The corresponding flow structures are illustrated qualitatively in Figure 10.1. In the generic configuration, the jet is injected perpendicularly to the main flow. However, variations of the injection direction with an angle against (partial counterflow) or in (partial coflow) the main flow direction are possible. Indeed such designs can be considered to help against the risk of flameholding. The jet cross-section, which is circular in the generic configuration, may have a different shape (e.g., elliptical) for achieving a certain effect. In arrays of multiple jets, if the jets are close enough to interact, the flow characteristics are additionally affected by the configuration.

The jet penetration depth (the mean jet trajectory) is generally found to correlate with the jet-to-freestream momentum flux ratio J given by the following expression:

$$J = \frac{(\rho u^2)_J}{(\rho u^2)_\infty} \qquad (10.1)$$

Flashback Mechanisms in Lean Premixed Gas Turbine Combustion

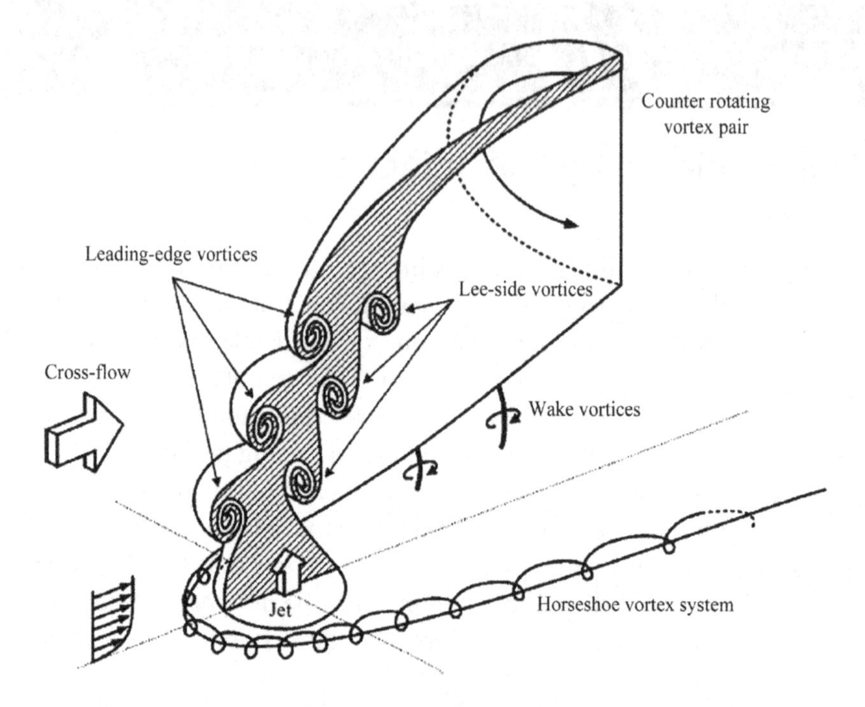

Fig. 10.1. Schematic of flow structures produced by jet in cross-flow (New et al., 2003).

where the subscripts J and ∞ denote the jet and the cross-flowing main stream, respectively, and u denotes the bulk speeds of the streams. For a given jet shape (normally circular) and injection angle (normally perpendicular to the freestream), correlations that describe the jet penetration have usually the following form (Ben-Yakar et al., 2006):

$$\frac{y}{dJ} = A\left(\frac{x}{dJ}\right)^m \tag{10.2}$$

In the previous equation, x denotes the distance measured along the freestream direction, y the jet penetration distance measured in the traversal direction, and d the jet diameter, while A and m denote empirical coefficients.

10.2 JET FLAMES IN CROSS-FLOW

An area where flames of jets in cross-flow play an important role has been rocket engines. Since H_2 is often used as fuel in the corresponding experiments, this experience may be seen to have certain relevance for the present purpose. In those applications the speeds are very high and the flow is normally supersonic. They exhibit quite strong shock waves, which interact with the turbulence and combustion and affect ignition and flameholding characteristics. The so-called "bow shock" on the wind side is typical for such jets. The strong pressure and temperature rise and accompanying flow deceleration behind the shock wave often cause autoignition and flame stabilization in this region, whereas in subsonic jets the flame may be extinguishing due to high strain. The main design target of these applications is contrary to the present interest, i.e., flameholding. In such investigations (Ben-Yakar and Hanson, 1998), the resulting strong upstream and downstream wall flow separations, the recirculation zones, are reported to be welcome, as they provide regions for radical production and enhance flameholding.

Thus, for the present purposes (prevention of flameholding) the recirculation regions in front (horseshoe vortex) and behind the jet exit should be minimized. A slender jet shape aligned with the flow (e.g., elliptical) was observed to create smaller recirculation zones, and thus reduce flameholding capability of the jet. Partial coflow injection was also found to reduce the flameholding capability of the jets.

Ben-Yakar et al. (2006) performed measurements on hydrogen and ethylene jets injected into a supersonic cross-flow. An interesting outcome of the study was that the penetration depths for hydrogen and ethylene jets were different, although the jet-to-freestream momentum flux ratio J was the same. This was attributed to an additional influence of the jet shear layer development, as the velocities of both jets were different significantly due to the different densities. It was found that the correlation of Rothstein and Wantuck (Ben-Yakar et al., 2006) provided a good prediction of the penetration depth of hydrogen and ethylene jets. In their correlation, the coefficient A in Eq. (10.2) is not a constant but a function of J, i.e., $A = 2.1773 / J^{0.443}$, and $m = 0.281$. Investigations performed also on subsonic jet flames are discussed in the following.

Kalghatgi (1981) performed a systematic experimental investigation on blowout stability of diffusion flames in cross-wind for different hydrocarbon fuels such as methane, propane, ethylene, and commercial butanes (but not for hydrogen). For each burner diameter, for each gas, a blowout stability limit curve was drawn in the U_e–V plane (U_e, jet exit

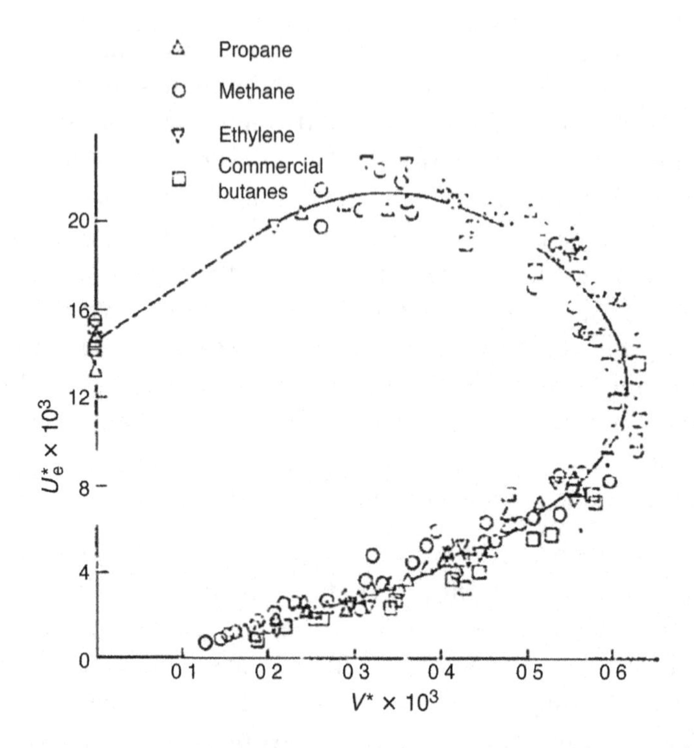

Fig. 10.2. The general stability curve of Kalghatgi (1981).

velocity; V, cross-wind velocity), which defines the boundaries of the region in which stable flames can exist. All data points could be made to collapse onto a general stability curve, shown in Figure 10.2, by plotting them in terms of the nondimensional jet exit (U_e^*) and cross-wind (V^*) velocities defined as follows:

$$U_e^* = \frac{U_e}{W}; \qquad V^* = \frac{V}{W} \tag{10.3}$$

where

$$W = S\left(\frac{H \cdot S}{\nu_e}\right)\left(\frac{\rho_\infty}{\rho_e}\right)^{1.5} \tag{10.4}$$

In Eq. (10.4), S is the maximum laminar flame speed for the fuel in a mixture with the ambient gas, ν_e and ρ_e are the kinematic viscosity and density of the fuel at the jet exit, ρ_∞ is the ambient gas density, and H is the distance along the burner axis at which the mean fuel concentration falls to its stoichiometric value in still air.

Thus, according to Kalghatgi (1981), for a given moderate cross-wind, there are two blowout limits – a stable lifted flame bent by the cross-wind can exist between these limits. Below the lower limit, the jet velocity is much smaller than the blowout limit for the case without cross-flow, and the flame is blown off by the wind. However, it was found that in some cases, usually when the burner diameter is large and V is small and the jet exhibits a rim, this lower limit may not exist. The flame gets stabilized in the wake of the fuel pipe, attaches itself to the pipe, and cannot be extinguished, even when U_e approaches zero. At the upper stability limit, the mechanism of blowout is similar to that of the flame in still air. It can be seen from Figure 10.2 that this upper limit is usually higher than that in the absence of cross-wind. The explanation for this may be found in the fact that, in general, the intensity of turbulence is considerably larger in a jet in cross-wind than in a jet in still air. Stability limits expand, with increasing burner diameter, which can be deduced from Eq. (10.4) as H scales with d. In a further study, Kalghatgi (1982) investigated the effect of injection angle. Partially coflowing configurations were found to be less stable.

Menon and Gollahalli (1988) investigated multiple jets in cross-flow, experimentally. Figure 10.3 shows the plots of jet velocity at blowout with cross-flow velocity for single- and multiple-jet flames with different diameters. The effect of the separation distance can also be seen in the figure.

The single-jet stability curve (Figure 10.3, $N = 1$, curves A and B) is similar to that of Kalghatgi (1981). This curve was obtained by keeping the jet velocity constant at a value, while increasing the cross-flow velocity. The upper part of the stability curve (branch A) could also be determined by keeping cross-flow velocity at the desired value and increasing jet velocity. However, when this procedure was applied to reach the conditions corresponding to the lower part of the curve (branch B), a third limit of jet velocity (branch C) was found at which the flame blows out. At this limit, the flame lifts off the burner and blows out immediately. At this limit, liftoff and blowout occur simultaneously. The comparably much higher cross-flow velocity leads to too lean a mixture.

One can see (Figure 10.3) that multiple-jet flames (with $a/d = 8$) are more stable than single-jet flames. For a given separation distance, the stability range increases as the number of jets increases. The merged

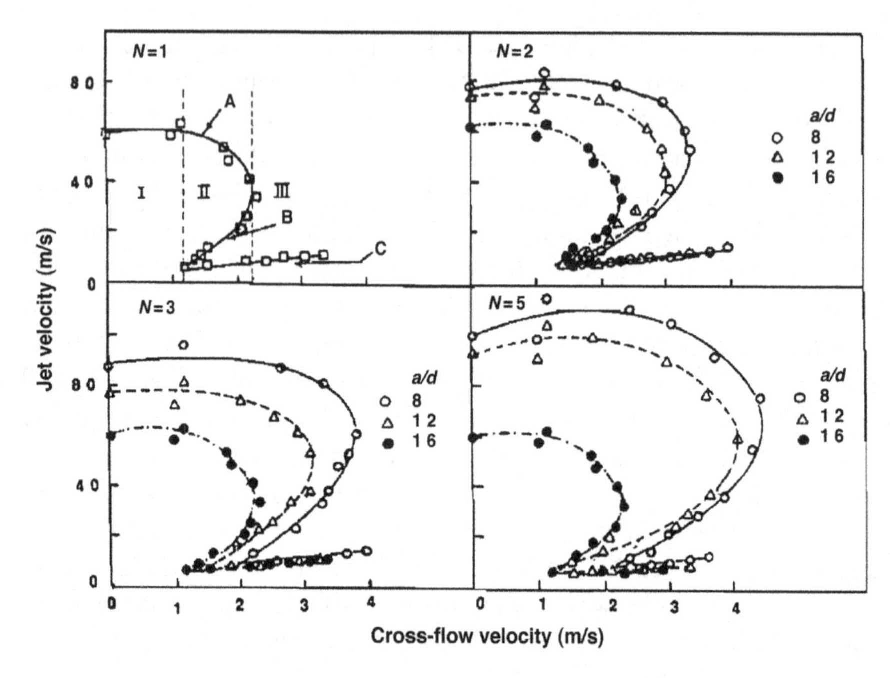

*Fig. 10.3. Stability curves of multiple jet diffusion flames in cross-flow: effect of separation distance (*a*) and number of jets (*N*) on flame blowout, for* d *= 1.8 mm (*N *= 1: single jet;* N *= 2, two jets side by side;* N *= 3, three jets side by side;* N *= 5: five jets arranged as a cross;* d*: jet diameter,* a*: separation distance between jets) (Menon and Gollahalli, 1988).*

flame behaves as the flame of a larger diameter jet, leading to larger blowout jet velocities. The stability range of jet velocity decreases with increasing separation distance, in all configurations. As separation distance is increased, the distance at which the jets merge also increases. This means increased air entrainment and leaner conditions at the base of the merged flame. Consequently, the flame blows out at a smaller jet velocity. One can also see (Figure 10.3) that the flame-to-flame interaction is no more significant and the jets behave as single entities when the separation distance reaches about 16 jet diameters. An interesting point is that the branch C of the stability curve is practically independent of the number of jets or their configuration. The difference between the stability curves $N = 2$ and $N = 3$ cases at a given separation distance is not substantial. The stability curve for the $N = 5$ case is, however, much wider. This can be attributed to the additional interaction of jets aligned with the cross-flow, as the upstream jet shields the downstream one from cross-flow, the downstream flame acting as a pilot ignitor for the upstream one. In case of the extinction of the upstream flame, its fuel gases can be reignited by the downstream one. Thus, aligning the jets parallel to cross-flow increases their resistance against blowout.

Hasselbrink and Mungal (1998) investigated lifted nonpremixed flames of methane jets in a cross-flow of air. Near blowout the flame was lifted and the combustion took place in a partially premixed mode. It was observed that low-velocity regions created by jet and cross-flow interaction play key roles in the flame stabilization. In particular, the lee-side flame base is in a strong recirculation zone behind the main jet and serves as an aerodynamic stabilizer. As the size of the recirculation zones scales with the jet diameter, one can conclude that jets with smaller diameter are less prone to flameholding.

Some results for the flameholding behavior of jets in cross-flow using hydrogen blend fuels were published in a report of GE Energy (2005). Some conclusions drawn were: (1) a mixture of 60% H_2 and 40% N_2 (volume fractions) was observed to have a low tolerance to flameholding (0.159 in. injector) compared with methane (0.5 in. injector). For methane, flameholding was not observed for air velocities above 200 fps. For hydrogen (60% H_2) air velocities around 400 fps were required to prevent flameholding, corresponding to about 1% in combustor pressure drop. (2) The effect of injection angle (+45°, 90°, −45°) on flameholding margin was found to be negligible.

Grout et al. (2011) performed DNS of flame stabilization downstream of a transverse fuel jet in cross-flow. The core of the heat release was found to be located near the trailing edge of the fuel jet, at approximately four nozzle diameters away from the wall, and was characterized by the simultaneous occurrence of locally stoichiometric reactants and low flow velocities in the mean. The location where the most upstream tendrils of the flame were found was in the region where coherent vortical structures, originating from the jet shear layer interaction, were present. Instantaneously, upstream flame movement was observed through propagation into the outer layers of jet vortices.

Kolla et al. (2012) investigated mechanisms of flame stabilization and blowout in reacting turbulent hydrogen jet in cross-flow, numerically by DNS. The jet fuel consisted of 70% H_2 and 30% N_2 (v/v). The freestream velocity of the cross-flow was 56.5 m/s. The jet Reynolds number was 4000. The resulting velocity and momentum flux ratios were 4.5 and

3.6, respectively. For the isothermal flow, the investigated jet angles were 90° and 70° (angled in the flow direction). As the injection angle was reduced from 90° to 70°, the low-velocity region behind the jet was found to diminish significantly, in terms of both physical extent and magnitude, indicating reduced ability to provide favorable conditions for flame anchoring and stabilization. For the reacting case, jet angles 90° and 75° (partial coflow) were considered. As the injection angle is smoothly changed to 75° a transient flame blowout is observed. The sequence of instantaneous snapshots of temperature on the midplane displayed in Figure 10.4 shows that the flame blows out in the 75° simulation.

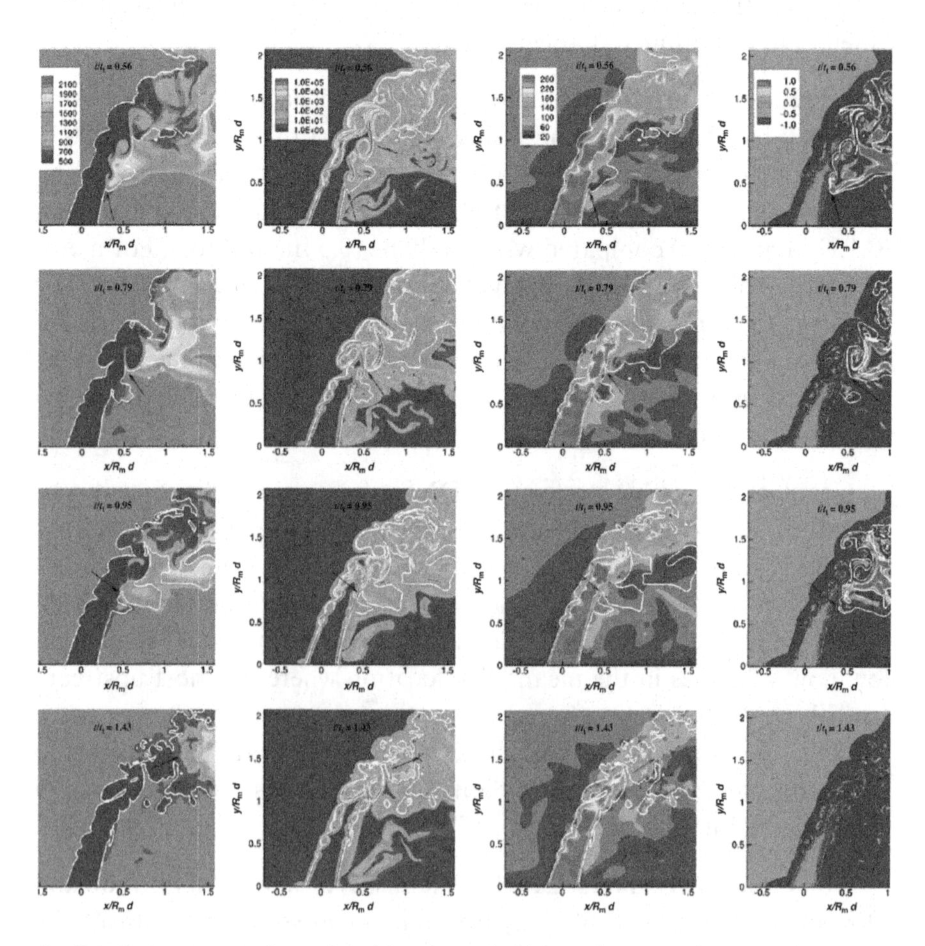

Fig. 10.4. Instantaneous snapshots on the midplane showing from left to right contours of temperature, mixture fraction scalar dissipation rate, velocity magnitude, and flame index. The white line is the isocontour of stoichiometric mixture fraction in the left three columns, and in the right column represents 10% and 90% of peak heat release rate. The arrow indicates the flame root location (Kolla et al., 2012).

The simulations (Kolla et al., 2012) were found to strengthen the hypothesis that the conditions critical for flame stabilization are the presence of near-stoichiometric mixtures in regions of low velocities and scalar dissipation rates. The results indicated two phases of blowout. In the first phase, dominant flow structures such as shear layer vortices entrain the flame base closer to the jet centerline. Thus, the flame base encounters rich mixtures and higher velocities and the imbalance between flame speed and flow velocity initiates flame blowout. As the flame base moves downstream, the second phase starts during which, in spite of low velocities, low scalar dissipation rates, and near-stoichiometric mixtures, the flow velocity normal to the flame base increases quickly. Due to jet bending, the flame base becomes increasingly normal to the cross-flow maintaining the blowout.

For jets in cross-flow, a few parameters that influence the flameholding behavior are summarized as follows:

- Jet exit diameter: Most of the experiments show a clear correlation between the jet diameter and flameholding propensity. Larger jet diameter means larger recirculation zones around the jet exit that act as aerodynamic flameholders. Thus, for reducing the flameholding propensity jet diameter should be reduced.
- Jet exit aspect ratio: Jets with noncircular, slender, such as elliptical, cross-sections aligned with the cross-flow direction have a reduced flameholding ability, compared with a round jet.
- Jet injection angle: In some experiments, a partial coflow injection angle was observed to reduce the flameholding ability compared to perpendicular injection.
- Jet spacing (for multiple jets): Multiple jets generally show a stronger resistance against blowout compared to a single jet. A jet row aligned with the freestream direction has a larger flameholding capability compared with jets arranged perpendicular to the freestream direction. For spacing distance larger than 16 jet diameters, the interaction becomes negligible and the jets behave as single jets.
- Jet composition: Higher H_2 content increases the flameholding propensity.

REFERENCES

Abdel-Gayed, R.G., Bradley, D., 1989. Combustion regimes and the straining of the turbulent premixed flames. Combust. Flame 76, 213–218.

Abdel-Gayed, R.G., Bradley, D., Hamid, M.N., Lawes, M., 1984. Lewis number effects on turbulent burning velocity. Proc. Combust. Inst. 20, 505–512.

Asato, K., Wada, H., Hiruma, T., 1997. Characteristics of flame propagation in a vortex core: validity of a model for flame propagation. Combust. Flame 110, 418–428.

Ashurst, W.M.T., 1996. Flame propagation along a vortex: the baroclinic push. Combust. Sci. Technol. 112, 175–185.

Baumgartner, G., Sattelmayer, T., 2013. Experimental investigation of the flashback limits and flame propagation mechanisms for premixed hydrogen–air flames in non-swirling and swirling flow. In: ASME Paper GT2013-94258.

Beerer, D.J., 2009. Autoignition of Methane, Ethane, Propane and Hydrogen in Turbulent High Pressure and Temperature Flows. M.Sc. Thesis. Mechanical and Aerospace Engineering, University of California, Irvine CA, USA.

Beerer, D.J., McDonnel, V.G., 2008. Autoignition of hydrogen and air inside a continuous flow reactor with application to lean premixed combustion. J. Eng. Gas Turbines Power 130, 051507-1–051507-8.

Ben-Yakar, A., Hanson, R.K., 1998. Experimental investigation of flame-holding capability of hydrogen transverse jet in supersonic cross-flow. Proc. Combust. Inst. 27, 2173–2180.

Ben-Yakar, A., Mungal, M.G., Hanson, R.K., 2006. Time evolution and mixing characteristics of hydrogen and ethylene transverse jets in supersonic crossflows. Phys. Fluids 18, 026101-1–026101-16.

Benjamin, B., 1962. Theory of the vortex breakdown phenomenon. J. Fluid Mech. 14, 593–629.

Blesinger, G., Koch, R., Bauer, H.-J., 2010. Influence of flow field scaling on flashback of swirl flames. Exp. Thermal Fluid Sci. 34, 290–298.

Blumenthal, R., Fieweger, K., Komp, H., Adomeit, G., 1996. Gas dynamic features of self ignition of non-diluted fuel/air mixtures at high pressure. Combust. Sci. Technol. 113–114, 137–166.

Boger, M., Veynante, D., Boughanem, H., Trouve, A., 1998. Direct numerical simulation analysis of the flame surface density concept. Proc. Combust. Inst. 27, 917–925.

Bollinger, L.E., 1958. Evaluation of flame stability at high Reynolds numbers. Jet Propulsion 28, 334–335.

Bollinger, L.E., Edse, R., 1956. Effect of burner-tip temperature on flash back of turbulent hydrogen–oxygen flames. Ind. Eng. Chem. 48, 802–807.

Borghi, R., 1988. On the structure and morphology of turbulent premixed flames. Recent Advances in the Aerospace Science. Pergamon Press, New York, pp. 117–139.

Bradley, D., 1992. How fast can we burn? Proc. Combust. Inst. 24, 247–253.

Brandl, A., Mooney, J.D., Pfitzner, M., Durst, B., Kern, W., 2005. Comparison of combustion models and assessment of their applicability to the simulation of premixed turbulent combustion in IC-engines. Flow Turbulence Combust. 75, 335–350.

Brown, G., Lopez, J., 1990. Axisymmetric vortex breakdown. Part 2: physical mechanisms. J. Fluid Mech. 221, 553–576.

Flashback Mechanisms in Lean Premixed Gas Turbine Combustion

Catlin, C.A., Lindstedt, R.P., 1991. Premixed turbulent burning velocities derived from mixing controlled reaction models with cold front quenching. Combust. Flame 85, 427–439.

Cheng, R.K., Littlejohn, D., Strakey, P.A., Sidwell, T., 2009. Laboratory investigations of a low-swirl injector with H_2 and CH_4 at gas turbine conditions. Proc. Combust. Inst. 32, 3001–3009.

Cheng, R.K., Oppenheim, A.K., 1984. Auto-ignition in methane hydrogen mixtures. Combust. Flame 58, 125–139.

Chomiak, J., 1976. Dissipation fluctuations and the structure and propagation of turbulent flames in premixed gases at high Reynolds numbers. Proc. Combust. Inst. 16, 1665–1673.

Coats, C.M., 1980. Comment on review of flashback reported in prevaporizing/premixing combustors. Combust. Flame 37, 331–333, (Authors' reply, 335–336).

Cohé, C., Halter, F., Chauveau, C., Gökalp, I., Gülder, Ö., 2007. Fractal characterization of high-pressure and hydrogen-enriched CH_4–air turbulent premixed flames. Proc. Combust. Inst. 31, 1345–1352.

Colin, P., Ducros, F., Veynante, D., Poinsot, D., 2000. A thickened flame model for large eddy simulations of turbulent premixed combustion. Phys. Fluids 12, 1843–1863.

Cowell, L., Lefebvre, A., 1986. Influence of pressure on autoignition characteristics of gaseous hydrocarbon–air mixtures. In: SAE Technical Paper 860068.

Dam, B., Corona, G., Hayder, M., Choudhuri, A., 2011a. Effects of syngas composition on combustion induced vortex breakdown (CIVB) flashback in a swirl stabilized combustor. Fuel 90, 3274–3284.

Dam, B., Love, N., Choudhuri, A., 2011b. Flashback propensity of syngas fuels. Fuel 90, 618–625.

Daneshyar, H., Hill, P.G., 1987. The structure of small-scale turbulence and its effect on combustion in spark ignition engines. Prog. Energy Combust. Sci. 13, 47–73.

Daniele, S., Jansohn, P., 2012. Correlations for turbulent flame speed of different syngas mixtures at high pressure and temperature. In: ASME Paper GT2012-69611.

Davis, S.G., Joshi, A.V., Wang, H., Egolfopoulos, F., 2005. An optimized kinetic model of H_2/CO combustion. Proc. Combust. Inst. 30, 1283–1292.

De, A., Acharya, S., 2012. Dynamics of upstream flame propagation in a hydrogen enriched premixed flame. Int. J. Hydrogen Energy 37, 17294–17309.

DeVries, J., Petersen, E., 2007. Autoignition of methane based fuel blends under gas turbine conditions. Proc. Combust. Inst. 31, 3163–3171.

Di Sarli, V., Di Benedetto, A., 2007. Laminar burning velocity of hydrogen methane/air premixed flames. Int. J. Hydrogen Energy 32, 637–646.

Dinkelacker, F., Manickam, B., Muppala, S.P.R., 2011. Modelling and simulation of lean premixed turbulent methane/hydrogen/air flames with an effective Lewis number approach. Combust. Flame 158, 1742–1749.

Driscoll, J.F., 2008. Turbulent premixed combustion: flamelet structure and its effect on turbulent burning velocities. Prog. Energy Combust. Sci. 34, 91–134.

Durbin, P.A., Reif, B.A.P., 2010. Statistical Theory and Modeling for Turbulent Flows, second ed. Wiley, New York.

Eichler, C.T., 2011. Flame Flashback in Wall Boundary Layers of Premixed Combustion Systems. Dissertation. Technical University of Munich, Munich, Germany.

Eichler, C., Baumgartner, G., Sattelmayer, T., 2011. Experimental investigation of turbulent boundary layer flashback limits for premixed hydrogen–air flames confined in ducts. In: ASME Paper GT2011-45362.

Eichler, C., Sattelmayer, T., 2010. Experiments on flame flashback in a quasi-2D turbulent wall boundary layer for premixed methane–hydrogen–air mixtures. In: ASME Paper GT2010-23401.

von Elbe, G., Mentser, M., 1945. Further studies of the structure and stability of burner flames. J. Chem. Phys. 13, 89–100.

Eroglu, A., Döbelling, K., Joos, F., Brunner, P., 2001. Vortex generators in lean premix combustion. J. Eng. Gas Turbines Power 123, 41–49.

Escudier, M.P., 1988. Vortex breakdown: observations and explanations. Prog. Aerospace Sci. 25, 189–229.

Fine, B., 1958. The flashback of laminar and turbulent burner flames at reduced pressure. Combust. Flame 2, 253–266.

Fine, B., 1959. Effect of initial temperature on flash back of laminar and turbulent flames. Ind. Eng. Chem. 51, 564–566.

Fritz, J., 2003. Flammenrückschlag durch verbrennungsinduziertes Wirbelaufplatzen. Dissertation. Technical University of Munich, Munich, Germany.

Fritz, J., Kröner, M., Sattelmayer, T., 2004. Flashback in a swirl burner with cylindrical premixing zone. J. Eng. Gas Turbine Power 126, 276–283.

GE Energy, 2005. Premixer design for high hydrogen fuels. In: Report Submitted to US Department of Energy, National Energy Technology Laboratory, Morgantown, WV. GE Energy Schenectady, New York. DOE Cooperative Agreement No. DE-FC26-03NT41893.

Glassman, J., Yetter, R.A., 2008. Combustion, fourth ed. Academic Press, Amsterdam.

Göttgens, J., Mauss, F., Peters, N., 1992. Analytic approximations of burning velocities and flame thicknesses of lean hydrogen, methane. Ethylene, ethane, acetylene, and propane flames. Proc. Combust. Inst. 24, 129–135.

Goy, C., Moran, A., Thomas, G., 2001. Autoignition characteristics of gaseous fuels at representative gas turbine conditions. In: ASME Paper 2001-GT-0051.

Grout, R.W., Gruber, A., Yoo, C.S., Chen, J.H., 2011. Direct numerical simulation of flame stabilization downstream of a transverse fuel jet in cross-flow. Proc. Combust. Inst. 33, 1629–1637.

Gülder, Ö.L., 1990. Turbulent premixed flame propagation models for different combustion regimes. Proc. Combust. Inst. 23, 734–750.

Guo, H., Tayebi, B., Galizzi, C., Escudié, D., 2010. Burning rates and surface characteristics of hydrogen enriched turbulent lean premixed methane–air flames. Int. J. Hydrogen Energy 35, 11342–11348.

Hasegawa, T., Michikami, S., Nomura, T., 2002. Flame development along a straight vortex. Combust. Flame 129, 294–304.

Hasegawa, T., Noguchi, S., 1997. Numerical study of a turbulent flow compressed by a weak shock wave. Int. J. Comput. Fluid Dyn. 8, 63–75.

Hasselbrink, E.F., Mungal, M.G., 1998. Observations on the stabilization region of lifted non-premixed methane transverse jet flames. Proc. Combust. Inst. 27, 1167–1173.

Hawkes, E.R., Chen, J.H., 2004. Direct numerical simulation of hydrogen-enriched lean premixed methane–air flames. Combust. Flame 138, 242–258.

Huang, J., Hills, P., Bushe, K., Munshi, S., 2004. Shock-tube study of methane ignition under engine relevant conditions: experiment and modeling. Combust. Flame 136, 25–42.

Huang, Z., Zhang, Y., Zeng, K., Liu, B., Wang, Q., Jiang, D., 2006. Measurements of laminar burning velocities for natural gas–hydrogen–air mixtures. Combust. Flame 146, 302–311.

Ishizuka, S., 2002. Flame propagation along a vortex axis. Prog. Energy Combust. Sci. 28, 477–542.

Ishizuka, S., Hamasaki, T., Koumura, K., Hasegawa, R., 1998. Measurements of flame speeds in combustible vortex rings: validity of the back-pressure drive flame propagation mechanism. Proc. Combust. Inst. 27, 727–734.

Kalghatgi, G.T., 1981. Blow-out stability of gaseous jet diffusion flames, part II: effect of cross wind. Combust. Sci. Technol. 26, 241–244.

Kalghatgi, G.T., 1982. Blow-out stability of gaseous jet diffusion flames, part III: effect of burner orientation to wind direction. Combust. Sci. Technol. 28, 241–245.

Karpov, V.P., Lipatnikov, A.N., Wolanski, P., 1997. Finding the Markstein number using the measurements of expanding spherical laminar flames. Combust. Flame 109, 436–448.

Keller, J.J., 1995. On the interpretation of vortex breakdown. Phys. Fluids 7, 1695–1702.

Keller, J.O., Vaneveld, K., Korschelt, D., Hubbard, G.L., Ghoniem, A.F., Daily, J.W., Oppenheim, A.K., 1982. Mechanism of instabilities in turbulent combustion leading to flashback. AIAA J. 20, 254–262.

Khitrin, L.N., Moin, P.B., Smirnov, D.B., Shevchuk, V.U., 1965. Peculiarities of laminar- and turbulent-flame flashbacks. Proc. Combust. Inst. 10, 1285–1291.

Kido, H., Nakahara, M., Nakashima, K., Hashimoto, J., 2002. Influence of local flame displacement velocity on turbulent flame velocity. Proc. Combust. Inst. 29, 1855–1861.

Kieswetter, F., Konle, M., Sattelmayer, T., 2007. Analysis of combustion induced vortex breakdown driven flame flashback in a premix burner with cylindrical mixing zone. J. Eng. Gas Turbines Power 129, 929–936.

Kobayashi, H., Kawabata, Y., Maruta, K., 1988. Experimental study on general correlation of turbulent burning velocity at high pressure. Proc. Combust. Inst. 27, 941–948.

Kobayashi, H., Tamura, T., Maruta, K., Niioka, T., Williams, F.A., 1996. Burning velocity of turbulent premixed flames in a high pressure environment. Proc. Combust. Inst. 26, 389–396.

Kolla, H., Grout, R.W., Gruber, A., Chen, J.H., 2012. Mechanisms of flame stabilization and blowout in reacting turbulent hydrogen jet in cross-flow. Combust. Flame 159, 2755–2766.

Konle, M., Sattelmayer, T., 2010. Time scale model for the prediction of the onset of flame flashback driven by combustion induced vortex breakdown. J. Gas Turbines Eng. Power 132, 041503-1–041503-6.

Krebs, W., Hellat, J., Eroglu, A., 2010. Technische Verbrennungssysteme. In: Lechner, C., Seume, J. (Eds.), Stationäre Gasturbinen. second revised ed. Springer-Verlag, Berlin, pp. 453–490, (Chapter 10).

Kröner, M., 2003. Einfluss lokaler Löschvorgänge auf den Flammenrückschlag durch verbrennungsinduziertes Wirbelaufplatzen. Dissertation. Technical University of Munich, Munich, Germany.

Kröner, M., Sattelmayer, T., Jassin, F., Kieswetter, F., Hirsch, Ch., 2007. Flame propagation in swirling flows – effect of local extinction on the combustion induced vortex breakdown. Combust. Sci. Technol. 179, 1385–1416.

Law, C.L., 2006. Combustion Physics. Cambridge Press, Cambridge.

Law, C.L., Kwon, O.C., 2004. Effects of hydrocarbon substitution on atmospheric hydrogen–air flame propagation. Int. J. Hydrogen Energy 29, 867–879.

Lefebvre, A.H., 1983. Gas Turbine Combustion. Hemisphere Publishing Corporation, New York.

Lewis, B., von Elbe, G., 1987. Combustion, Flames and Explosion of Gases. Academic Press, London.

Li, S.C., Williams, F.A., 2002. Reaction mechanisms for methane ignition. J. Eng. Gas Turbines Power 124, 471–480.

Libby, P.A., Williams, F.A. (Eds.), 1994. Turbulent Reacting Flows. Academic Press, London.

Lieuwen, T., 2008. Flashback characteristics of syngas-type fuels under steady and pulsating conditions. In: Final Report. DOE Award Number: DE-FG26-04NT42176. School of Aerospace Engineering, Georgia Institute of Technology, Atlanta, GA.

Lieuwen, T., McDonnel, V., Petersen, E., Santavicca, D., 2008a. Fuel flexibility influences on premixed combustion blowout, flashback, autoignition, and stability. J. Eng. Gas Turbines Power 130, 011506-1–011506-10.

Lieuwen, T., McDonnel, V., Santavicca, D., Sattelmayer, T., 2008b. Burner development and operability issues associated with steady flowing syngas fired combustors. Combust. Sci. Technol. 180, 1169–1192.

Lindstedt, R.P., Vaos, E.M., 1999. Modeling of turbulent flames with second moment methods. Combust. Flame 116, 461–485.

Lipatnikov, A.N., Chomiak, J., 2002. Turbulent flame speed and thickness: phenomenology, evaluation and application in multi-dimensional simulations. Prog. Energy Combust. Sci. 18, 1–74.

Liu, Y., 1991. Untersuchung zur stationären Ausbreitung turbulenter Vormischflammen. Dissertation. University of Karlsruhe, Karlsruhe, Germany.

Liu, Y., Lenze, B., 1988. The influence of turbulence on the burning velocity of premixed CH_4–H_2 flames with different laminar burning velocities. Proc. Combust. Inst. 22, 747–754.

Lucca-Negro, O., O'Doherty, T., 2001. Vortex breakdown: a review. Prog. Energy Combust. Sci. 27, 431–481.

Mandilas, C., Ormsby, M.P., Sheppard, C.G.W., Woolley, R., 2007. Effects of hydrogen addition on laminar and turbulent premixed methane and iso-octane air flames. Proc. Combust. Inst. 31, 1443–1450.

Mantel, T., Egolfopoulos, F.N., Bowman, C.T., 1996. A new methodology to determine kinetic parameters for one- and two-step chemical models. In: Proceedings of the Summer Program 1996. Centre for Turbulence Research. Stanford University, Stanford, CA, USA pp. 149-166.

Mashuga, C.V., Crowl, D.A., 2000. Derivation of Le Chatelier's mixing rule for flammable limits. Process Saf. Prog. 19, 112–117.

Mayer, C., Sangl, J., Sattelmayer, T., Lachaux, T., Bernero, S., 2011. Study on the operational window of a swirl stabilized syngas burner under atmospheric and high pressure conditions. In: ASME Paper GT2011-45125.

McCormack, P.D., Scheller, K., Mueller, G., Tisher, R., 1972. Flame propagation in a vortex core. Combust. Flame 19, 297–303.

Meneveau, C., Poinsot, T., 1991. Stretching an quenching of flamelets in premixed turbulent combustion. Combust. Flame 86, 311–322.

Menon, R., Gollahalli, S.R., 1988. Combustion characteristics of interaction multiple jets in cross flow. Combust. Sci. Technol. 60, 375–389.

Meyer, J., Oppenheim, A.K., 1971. On the shock-induced ignition of explosive gases. Proc. Combust. Inst. 13, 1153–1164.

Mittal, G., Sung, C.-J., Yetter, R.A., 2006. Autoignition of H2/CO at elevated pressures in a rapid compression machine. Int. J. Chem. Kinet. 38, 516–529.

Mueller, M.A., Kim, T.J., Yetter, R.A., Dryer, F.L., 1999a. Flow reactor studies and kinetic modeling of the H_2/O_2 reaction. Int. J. Chem. Kinet. 31, 113–125.

Mueller, M.A., Yetter, R.A., Dryer, F.L., 1999b. Flow reactor studies and kinetic modeling of the $H_2/O_2/NO_X$ and $CO/H_2O/O_2/NO_X$ reactions. Int. J. Chem. Kinet. 31, 705–724.

Munson, B.R., Young, D.F., Okiishi, T.H., Huebsch, W.W., 2009. Fundamentals of Fluid Mechanics, sixth ed. Wiley, New Jersey.

Najm, H.N., Ghoniem, A.F., 1994. Coupling between vorticity and pressure oscillations in combustion instability. J. Propulsion Power 10, 769–776.

New, T.H., Lim, T., Luo, S.C., 2003. Elliptic jets in cross flow. J. Fluid Mech. 494, 119–140.

Noble, D.R., Zhang Q., Lieuwen, T., 2006a. Hydrogen effects upon flashback and blowout. In: Proceedings of ICEPAG2006, International Colloquium on Environmentally Preferred Advanced Power Generation, September 5-8, 2006, Newport Beach, California, ICEPAG 2006-24012.

Noble, D.R., Zhang Q., Shareef, A., Tootle, J., Meyers, A., Lieuwen, T., 2006b. Syngas mixture composition effects upon flashback and blowout. In: ASME Paper GT2006-90670.

O'Conaire, M., Curran, H., Simmie, J., Pitz, W., Westbrook, C., 2004. A comprehensive modeling study of hydrogen oxidation. Int. J. Chem. Kinet. 36, 603–622.

Peters, N., Rogg, B. (Eds.), 1993. Reduced Kinetic Mechanisms for Applications in Combustion Systems. Springer-Verlag, Berlin.

Petersen, E.L., Davidson, D.F., Hanson, R.K., 1999. Kinetics modeling of shock induced ignition in low-dilution CH_4/O_2 mixtures at high pressures and intermediate temperatures. Combust. Flame 117, 272–290.

Petersen, E.L., Röhrig, M., Davidson, D.F., Hanson, R.K., Bowman, C.T., 1996. High pressure methane oxidation behind reflected shock waves. Proc. Combust. Inst. 26, 799–806.

Plee, S.L., Mellor, A.M., 1978. Review of flashback reported in prevaporizing/premixing combustors. Combust. Flame 32, 193–203.

Poinsot, T., Candel, S., Trouve, A., 1995. Application of direct numerical simulation to premixed turbulent combustion. Prog. Energy Combust. Sci. 21, 531–576.

Sabia, P., Schiesswohl, E., de Joannon, M.R., Cavaliere, A., 2006. Numerical analysis of hydrogen mild combustion. Turk. J. Eng. Environ. Sci. 30, 127–134.

Sagaut, P., 2006. Large Eddy Simulation for Incompressible Flows, third ed. Springer, Berlin.

Sarpkaya, T., 1971. On stationary and travelling vortex breakdowns. J. Fluid Mech. 45, 545–559.

Sattelmayer, T., 2003. Influence of the combustor aerodynamics on combustion instabilities from equivalence ratio fluctuations. J. Eng. Gas Turbines Power 125, 11–19.

Sattelmayer, T., 2010. Grundlagen der Verbrennung in stationären Gasturbinen. In: Lechner, C., Seume, J. (Eds.), Stationäre Gasturbinen. second revised ed. Springer-Verlag, Berlin, pp. 397–452, (Chapter 9).

Sattelmayer, T., Mayer C., Sangl, J., 2014. Interaction of flame flashback mechanisms in premixed hydrogen–air swirl flames. In: ASME Paper GT2014-25553.

Schefer, R.W., White, C., Keller, J., 2008. Lean hydrogen combustion. In: Dunn-Rankin, D. (Ed.), Lean Combustion Technology and Control. Academic Press, London.

Schlichting, H., 1979. Boundary Layer Theory, seventh ed. McGraw-Hill, New York.

Schmid, H.-P., Habisreuther, P., Leuckel, W., 1998. A model for calculating heat release in premixed turbulent flames. Combust. Flame 113, 79–91.

Shaffer, N., 2012. Personal communication.

Shaffer, N., Duan, Z., McDonnel, V., 2012. Study of fuel composition effects on flashback using a confined jet flame burner. In: ASME Paper GT2012-68401.

Sidwell, T., Richards, G., Casleton, K., Straub, D., Maloney, D., Strakey, P., Ferguson, D., Beer, S., Woodruff, S., 2006. Optically accessible pressurized research combustor for computational fluid dynamics model validation. AIAA J. 44, 434–443.

Spadaccini, L., Colket, M., 1994. Ignition delay characteristics of methane fuels. Prog. Energy Combust. Sci. 20, 431–460.

Squire, H.B., 1962. Analysis of the vortex breakdown phenomenon. In: Schaefer, M. (Ed.), Miszellaneen Der Angewandten Mechanik. Akademie Verlag, Berlin, pp. 306–312.

Strakey, P., Sidwell, T., Ontko, J., 2007. Investigation of the effects of hydrogen addition of lean extinction in a swirl stabilized combustor. Proc. Combust. Inst. 31, 3173–3180.

Ströhle, J., Myhvold, T., 2007. An evaluation of detailed reaction mechanisms for hydrogen combustion under gas turbine conditions. Int. J. Hydrogen Energy 32, 125–135.

Takemura, S., Umemura, A., 2002. Behaviors of a flame ignited by a hot spot in a combustable vortex (vortex-bursting initiation revisited). Proc. Combust. Inst. 29, 1729–1736.

Tangermann, E., Pfitzner, M., 2009. Evaluation of combustion models for combustion-induced vortex breakdown. J. Turbulence 10, 1–21.

Tangermann, E., Pfitzner, M., Konle, M., Sattelmayer, T., 2010. Large-eddy simulation and experimental observation of combustion-induced vortex breakdown. Combust. Sci. Technol. 182, 505–516.

Thibaut, D., Candel, S., 1998. Numerical study of unsteady premixed combustion: application to flashback simulation. Combust. Flame 113, 53–65.

Tian, X., Xing, S., Cui, Y., Fang, A., Nie, C., 2014. Numerical investigation of CIVB flashback process in a swirl-premixed burner with diverging centerbody. In: ASME Paper GT2014-25139.

Tsuobi, T., 1975. Homogeneous thermal oxidation of methane in reflected shock waves. Proc. Combust. Inst. 15, 883–890.

Tuncer, O., Acharya, S., Ulm, J.H., 2006. Hydrogen enriched confined methane flame behavior and flashback modeling. In: Forty-Fourth AIAA Aerospace Sciences Meeting and Exhibit, Reno, Nevada. AIAA Paper No: AIAA 2006-754.

Tuncer, O., Acharya, S., Ulm, J.H., 2009. Dynamics, NOx and flashback characteristics of confined premixed hydrogen-enriched methane flames. Int. J. Hydrogen Energy 34, 496–506.

Turns, S.R., 2012. An Introduction to Combustion, third ed. McGraw-Hill, Boston.

Umemura, A., Takamori, S., 2000. Wave nature in vortex-bursting initiation. Proc. Combust. Inst. 28, 1941–1948.

Umemura, A., Tomita, K., 2001. Rapid flame propagation in a vortex tube in perspective of vortex breakdown phenomena. Combust. Flame 125, 820–838.

Van den Schoor, F., Hermanns, R.T.E., van Oijen, J.A., Verplaetsen, F., de Goey, L.P.H., 2008. Comparison and evaluation of methods for the determination of flammability limits, applied to methane/hydrogen/air mixtures. J. Hazard. Mater. 150, 573–583.

Voevodski, V.V., Soloukhin, R.I., 1965. On the mechanism and explosion limits of hydrogen–oxygen chain self-ignition in shock waves. Proc. Combust. Inst. 10, 279–283.

Wang, S., Rusak, Z., 1997. The dynamics of a swirling flow in a pipe and transition to axisymmetric vortex breakdown. J. Fluid Mech. 340, 177–223.

Wankhede, M.J., Bressloff, N.W., Keane, A.J., Caracciolo, L., Zedda, M., 2010. An analysis of unstable flow dynamics and flashback mechanism inside a swirl stabilised lean burn combustor. In: ASME Paper GT2010-22253.

Westbrook, C.K., Dryer, F.L., 1981. Simplified reaction mechanism for the oxidation of hydrocarbon fuels in flames. Combust. Sci. Technol. 27, 31–43.

Williams, F., 1985. Combustion Theory, second ed. Addison Wesley, London.

Wohl, K., 1953. Quenching, flash-back, blow-off-theory and experiment. Proc. Combust. Inst. 4, 68–89.

Yetter, R.A., Rabitz, H., Hedges, R.M., 1991. A combined stability–sensitivity analysis of weak and strong reactions of hydrogen/oxygen mixtures. Int. J. Chem. Kinet. 23, 251–278.

Zimont, V.L., Biagioli, F., Syed, K., 2001. . Prog. Comput. Fluid Dyn. 1, 14–28.

NOMENCLATURE

A	flame surface area, constant
c_p	isobaric heat capacity
d	diameter
D	mass diffusivity
Da_T	Damköhler number
E_A	activation energy
g_C	critical velocity gradient
I	turbulence intensity
k_0	turbulence kinetic energy
K	flame stretch
Ka	Karlovitz number
l_0	integral length scale
L_M	Markstein length
Le	Lewis number
Ma	Markstein number
m	constant
n	overall reaction order, constant
Pr	Prandtl number
r	radial coordinate
r_C	vortex core radius
R	gas constant, radius
Re	Reynolds number
Re_T	turbulence Reynolds number
S	swirl number
S_L	laminar flame speed
S_T	turbulent flame speed
Sc	Schmidt number
t	time
T	temperature
T_B	temperature of burnt gas
T_{IG}	ignition temperature
T_U	temperature of unburned gas
u	velocity, axial velocity

u'	root mean square of velocity in turbulent flow
U	bulk axial velocity
U_f	flame tip propagation speed in CIVB
w	swirl (azimuthal, tangential) velocity
W_{max}	maximum swirl velocity at the edge of the vortex core
x	axial coordinate
X_i	mole fraction of species i
y	traversal coordinate, wall distance
y^+	nondimensional wall distance

Greek symbols

α	thermal diffusivity
δ_L	thickness of the laminar flame
δ_{PH}	thickness of the preheat zone
δ_R	thickness of the reaction zone
ε	dissipation rate of k
ϕ	equivalence ratio
η	Kolmogorov length scale
λ	thermal conductivity, Taylor length scale, excess air ratio
ν	kinematic viscosity
θ	azimuthal coordinate
ρ	density
ρ_b	density of burned mixture
ρ_u	density of unburned mixture
σ	unburned-to-burned density ratio
τ_C	chemical timescale
τ_{ig}	ignition delay time
τ_T	timescale of most energetic turbulence eddies
τ_W	wall shear stress
τ_η	Kolmogorov timescale
ω	vorticity
Ω	angular speed of vortex core

Abbreviations

BML	Bray–Moss–Libby
CFD	computational fluid dynamics
CIVB	combustion-induced VB

DNS	direct numerical simulations
FF	flame front
HS	high swirl
IRZ	internal recirculation zone
LES	large eddy simulations
LFL	lower flammability limit
LRR	Launder–Reece–Rodi
LS	low swirl
NTP	normal temperature and pressure
PIV	particle image velocimetry
PFR	plug flow reactor
PSR	perfectly stirred reactor
PVC	precessing vortex core
RANS	Reynolds averaged numerical simulations
RSM	differential Reynolds stress turbulence model
TBVA	turbulent burning along the vortex axis
UFL	upper flammability limit
URANS	unsteady RANS
VB	vortex breakdown
WBLF	wall boundary layer flashback

Printed in the United States
By Bookmasters